Agent-Based Social Systems

Volume 16

This series is intended to further the creation of the science of agent-based social systems, a field that is establishing itself as a transdisciplinary and cross-cultural science. The series will cover a broad spectrum of sciences, such as social systems theory, sociology, business administration, management information science, organization science, computational mathematical organization theory, economics, evolutionary economics, international political science, jurisprudence, policy science, socioinformation studies, cognitive science, artificial intelligence, complex adaptive systems theory, philosophy of science, and other related disciplines.

The series will provide a systematic study of the various new cross-cultural arenas of the human sciences. Such an approach has been successfully tried several times in the history of the modern science of humanities and systems and has helped to create such important conceptual frameworks and theories as cybernetics, synergetics, general systems theory, cognitive science, and complex adaptive systems.

We want to create a conceptual framework and design theory for socioeconomic systems of the twenty-first century in a cross-cultural and transdisciplinary context. For this purpose we plan to take an agent-based approach. Developed over the last decade, agent-based modeling is a new trend within the social sciences and is a child of the modern sciences of humanities and systems. In this series the term "agent-based" is used across a broad spectrum that includes not only the classical usage of the normative and rational agent but also an interpretive and subjective agent. We seek the antinomy of the macro and micro, subjective and rational, functional and structural, bottom-up and top-down, global and local, and structure and agency within the social sciences. Agent-based model-ing includes both sides of these opposites. "Agent" is our grounding for modeling; simula-tion, theory, and real-worldgrounding are also required.

As an approach, agent-based simulation is an important tool for the new experimental fields of the social sciences; it can be used to provide explanations and decision support for real-world problems, and its theories include both conceptual and mathematical ones. A conceptual approach is vital for creating new frameworks of the worldview, and the mathematical approach is essential to clarify the logical structure of any new framework or model. Exploration of several different ways of real-world grounding is required for this approach. Other issues to be considered in the series include the systems design of this century's global and local socioeconomic systems.

More information about this series at http://www.springer.com/series/7188

Toshiyuki Kaneda
Editor

Downtown Dynamics

Springer

Editor
Toshiyuki Kaneda
Graduate School of Engineering
Nagoya Institute of Technology
Nagoya, Aichi, Japan

ISSN 1861-0803 ISSN 2364-9542 (electronic)
Agent-Based Social Systems
ISBN 978-4-431-54900-0 ISBN 978-4-431-54901-7 (eBook)
https://doi.org/10.1007/978-4-431-54901-7

This Springer imprint is published by the registered company Springer Japan KK, part of Springer Nature.
The registered company address is: Shiroyama Trust Tower, 4-3-1 Toranomon, Minato-ku, Tokyo 105-6005, Japan

Preface

Downtown Dynamics for Agent-Based Urban Analytics

This book considers a series of exploratory studies based on the concept of a structural artificial society that contributes to the planning and management of crowded spaces in urban analytics research contexts. The current interests of the editor (who is experienced in urban modeling and simulation) involve exploring a composition of the research that consistently addresses data collection, analysis, modeling, and simulation for applications ranging from phenomenon analysis to urban space related policy exploration.

The downtown dynamics referred to in the title relate to the composition of an ABSS (Agent-Based Social Simulation) model, which introduces a spatiotemporal scale to manage the concept of crowding as both an urban phenomenon and an urban problem. To aim for dense grounding and policy intervention in the real world, the notion of middle-scale in the contexts of both urban scale and mesoscopic modeling is of great significance.

The crowding referred to in this book means the uneven time–space distribution of stay, flow, and other human activities in urban spaces. These are determined by the time–space distribution of the use of facilities where people's activities are conducted. Crowding in commercial districts may be a consequence of spatial behavior based primarily on individual economic motives; however, this not only embraces the efficiency of individual consumptive behavior but also includes many occurrences of spatial behavior (which are referred to as recreational time). This concept is not the same as economic activities in a narrow sense of the term simply because the activities occur within an urban space.

Furthermore, the spatial interests explored throughout this book are at the middle-scale, which at most is several times the size of the neighborhood unit. Until now, numerous downtown planning surveys and urban analysis studies concerning city planning have been conducted in terms of city center revitalization and location optimization plans. Despite providing an important grounding for the basis of each of these, there are many singular surveys, and the collection of data over time requires great expense and extensive research. In that sense, this domain has long been an underdeveloped field awaiting further exploration. Currently, systematic data acquisition via the IoT (Internet of Things) sensing is becoming possible in middle-scale urban spaces, and it is again attracting attention as a research frontier.

This book considers people who perform activities in an urban space and their interaction systems as a comprehensive multi-agent system; here, this system model has both a microscopic aspect (which focuses on the micro-factor actions of an individual) and a macroscopic aspect (which focuses on the macro-states of society to which each individual's planned actions and interaction results are aggregated).

In addition to this, the book establishes a mesoscope that focuses on local interaction fields for dense grounding and policy interventions, both of which are important requirements in urban analytics research. Here, an interaction field is a process into which individuals input their intentional (planned) behaviors and which outputs their realized behaviors. The interaction system has multiple interaction fields with relationships of differing distance to manage the immediate locality. The idea of such an agent-based approach (which models and simulates macro-social phenomena from the bottom-up based on micro-decision making in society among individuals) leads to the concept of artificial society, which is very simple in principle. Since the publication of Epstein and Axtell's 1996 book *Artificial Society*, the concept has become increasingly applied in current modeling and simulations.

This book undertook exploratory research to formulate answers to four specific questions. However, before addressing these questions, a summary of the Osu district of Nagoya city is presented; this location is the focus of a number of research surveys as it is a unique example of a downtown case study.

The Osu District: A Singular Study of a Lively Downtown District

The current Osu district is a 400 × 600 m shopping district comprising 1200 mainly small-scale shops located south of Wakamiya Odori Avenue in the midtown area of Nagoya city. This district, which combines numerous temples and commercial streets, can rightly be called the highlight of the Chukyo metropolitan area (with a population of around ten million in the Aichi, Gifu, and Mie prefectures). It constantly attracts diverse customers of all ages with its fascinating blend of old and new. This district features nine shopping streets arranged in a grid-like layout, each of which attracts certain customers by characteristically specific shop configurations. There are 30,000 visitors per day on weekdays and 70–80,000 at weekends and holidays; numbers of foreign tourists have also been increasing in recent years.

From the end of World War II until the 1970s, many shopping streets prospered in Japanese cities. When large-scale shopping malls started to appear and convenience stores gradually became popular, these traditional shopping streets fell into decay;

they were referred to as desolate streets around the year 2000 when the large-scale retail store law was abolished. In this situation, the district of Osu was one of the few shopping streets in Japan to survive, and its nostalgic atmosphere remains popular with shoppers.

Osu has a long history. Although it was built as a temple town around the many temples in the area (such as Osu Kannon and Banshoji) when Nagoya Castle was established in 1610, it retains the character of a post town due to the opening of Honmachi-Dori Street at the same time. It was also a licensed quarter for a time during the Meiji and Taisho eras, giving it the character of an entertainment district where shops, inns, playhouses, theaters, and cinemas were located. Later, in the 1930s, furniture stores started to appear in Uramonzencho making it the largest commercial area in Nagoya. Today, its shops sell a wide variety of goods, including clothing for middle-aged and elderly people, secondhand goods, and local products and souvenirs, and the area also offers numerous coffee shops.

During World War II, this district was damaged and was quickly reconstructed; however, it became inaccessible due to the newly established 100 m-wide Wakamiya Dori Street and the extension of Fushimi Dori Street in the reconstructed post-war city. Customer numbers fell due to the opening of city center department stores, the popularization of TV, and the delay in addressing the motorization so that by around 1970, the district was severely underpopulated.

However, since the mid-1970s, festival events held in collaboration with shop tenants and citizens and the growth in electronics parts stores have prompted visitor numbers to increase. The district continued to become more specialized as an electronics town that ranked alongside Tokyo's Akihabara or Osaka's Nipponbashi districts by around the year 2000. At this time, the arrival of clothes, food and drink, and entertainment shops for young people occurred. Later, the locations of PC and mobile phone shops were dispersed around the city center; however, sales were compensated by the emergence of subcultures such as otaku, manga, and cosplay, which comprised the more diverse attractions in this district. Furthermore, from around 2000, a variety of restaurants (such as those catering for Brazilian migrants) also became prominent, and Osu adopted a more diverse ethnic character.

Responsibility for the revitalization of Osu was taken by the young managers of some of the older shops in the district, and many festival events are held once every 10 days. Also, characteristic of the district is the rapid change in shop types, with vacant stores being quickly tenanted due to the endless stream of young people wishing to open their own retail premises.

Jane Jacobs' 1961 book *The Death and Life of Great American Cities* listed the conditions for city diversity, as follows: (1) (Two or More) Mixed primary uses, (2) (Not Too Large) Small blocks, (3) (Mix of Old and New) Aged buildings, and (4) Concentration (Density over the Threshold the Diversity Generates).

In contrast to Greenwich village during the era in which daily life was conducted within the community and also as a result of the expanding suburbanization of the Chukyo metropolitan area over the past 50 years, Osu has become a unique district supported by a metropolitan area of ten million people. Today, after the rise of both one-stop and online shopping, the downtown area is viewed by people as a recreational exploration destination. This stems from the desire for something more than just the functional efficiency of daily life.

Furthermore, if Osu is examined in light of the conditions for city diversity, the meaning of diversity and mixed use in the district is limited to commercial use. Excluding these points, it is evident that the insights from Jacobs' conditions for city diversity are not lost on this crowding research; rather, they provide important implications for this study. Here, it is noteworthy that of the four conditions for city diversity, (2) and (3) in particular are prescribed for small- to middle-scale urban spaces; they are indispensable components of urban space in the artificial society.

Downtown as a Phenomenon and the Visitors' Multi-purpose Multi-stop (MPMS) Behavior

The first question concerns the phenomenon known as downtown. In Japan, many shopping districts have fallen into decline over the last 50 years, but in Osu district in Nagoya the mainly small-scale shops and restaurants continue to attract the crowds. The downtown area has maintained this characteristic despite numerous environmental changes that appear to be part of a complex adaptive system; this is also an uneven and irregular phenomenon that causes change in the internal organization and external expression of the district.

Additionally, the district is a pool of diversity for the numerous varieties of shops and customers, and these pools are the source of adaptability for environmental and temporal changes. Despite the strong hypothesis that crowding in the Osu district is caused by diversity in commerce, how should research designs be specifically composed in order to both clarify these mechanisms and investigate the practical and effective policies for revitalizing downtown areas? Furthermore, within that question, what should the focus be for data collection?

To answer these questions, this book focuses on the shop-around behavior of visitors to the district. Shop-around behavior refers to the movement from shop to shop and between shops by visitors, and it is characterized by the concept of multi-purpose multi-stop (MPMS) behavior. It is a micro-behavioral unit, which implies an interaction between individual visitors and individual shops, and which provides positive externalities particularly with regard to these shops. Additionally, the district as a whole can be perceived as an accumulation of these activities.

Based on the above perspectives, the Kaneda lab at the Nagoya Institute of Technology conducted shop-around behavior surveys of visitors to the Osu district five times since 1998. In addition to collecting data on the shops visited, the purchase history, and the walking route data based on the network of the streets, this study's shop-around behavior surveys also asked questions on whether or not the actual visits to shops were planned, and the analysis of this data enables a decomposition of the behavior using the multilayered redundancy of visitors. Data regarding planned shop visits, in particular, provides a strong hint to assumptions of schedule planning, which is addressed later.

The Kaneda–Inagaki paper in Chap. 1 discusses the survey methods, basic analysis, spatial distributions of shop-around behavior, walking corridor patterns, and customer type analysis (in particular) of the shop-around behavior surveys and the shop type configuration surveys conducted in 1998, 2003, 2008, 2013, and 2018 in Osu. From the four surveys conducted since 2003, a kind of panel data was formed. Data collection was based entirely on questionnaire surveys and, even now, when it is possible to use big data based on sensing, the quality of each sample is high, and data that combines information about planned visits is useful despite the expense of sample costs.

In Chap. 2, the Kobayashi–Harazaki–Kaneda study provides a summary of the district based on the 2013 Osu survey, and it also provides a detailed factor analysis regarding shop tenant leases and closures in the 5-year period between 2008 and 2013. In addition to the strong correlation between the frequency of street usage and unplanned visits (per shop) to certain shop types, a strong correlation was found between the shop leave ratio and the street use frequency in the preceding 5 years among certain shop types.

The Takahashi–Kitazawa paper in Chap. 3 is a review of the research of visitors' spatial behavior on purchasing inside supermarkets and on in-store product layout. In supermarkets, purchasing data for each product category is accumulated and used as POS (Point Of Sales) data. Furthermore, in recent years IoT sensing has advanced by equipping visitors' shopping carts with transmission devices, such as RFID. Purchase data is collected at the point of sale, while the results are improved by means of analyses based on agent-based simulations. There is much to be gained from this research methodology, although the spatial scale is different to that in this study.

Spatial Maldistribution of Crowd and Visibility

The second question regards characteristic crowding factors in middle-scale urban spaces. Crowding is the uneven spatial distribution of activities. The two definitive factors of mass and distance are assessed by a gravity analogy model, which is mainly applied in large-scale urban spaces. What, however, are the new factors that should be considered as characteristic of middle-scale urban spaces in which dense grounding and policy intervention are both possible and inevitable?

In this book, the foremost candidates for the causes of crowding are spatial shape (in terms of road and block layout of the city) and visibility, which is the underlying principle. The analytical method used is a visibility graph analysis (VGA) based on computational geometry.

There have been various studies of urban morphology, but it has been difficult to target them using data analysis until now. The VGA was focused on as an indicator of formation as promoted by the Bartlett UCL School of Architecture and Planning.

VGA is a spatial morphologic measurement based on computational geometry. For that reason, it is essentially scale free, but this book focuses on its suitability as a tool for the downtown space scale, which has been the focus of research to date. All the VGA indicators have been developed to support the claim in the space syntax theory that people tend to gather where visibility is high. VGA is an analysis handling graph (network) with nodes on the plane and visibility links between them. It appeared after the 1990s when the computational approach advanced. VGA handles topological distances instead of physical distances, because it considers the shortest number of links between the nodes as a kind of distance measurement called the visual (step) depth.

Visibility can function as a new crowding factor, and it has the potential to secure grounding unique to the present when a large number of images and IoT sensing are socially widespread; therefore, it is expected that this will be an approach to new intervention.

Additionally, the business fields with an interest in this research topic at this scale of urban space include the real estate industry; for example, where it is common knowledge that rents at a ground floor commercial store are determined by the pedestrian flow volume on the road in front of the property. In fact, there is a very strong correlation between the two factors. This book focuses on land rent, property rent, and land price as a replacement index for crowding.

The research methodology addressed in the chapters is traditional, but for the (objective) crowding indicator variable, a multiple regression model with a combination of explanatory variables is prepared to provide the best fit. Comparisons are conducted on the strength of the standardized partial regression coefficient of each of the selected variables. This approach developed by Dr. Ota throughout the 2010s enables analysis by selecting the minimum value in Akaike's information criteria in information volume statistics. All three of the chapters in this book adopt this methodology, and it is expected that a meta-analysis will be conducted in the future.

In the Ota–Mizuno–Uriel–Kaneda paper in Chap. 4, a factor analysis is conducted targeting the 83,000 m^2 Sakae underground shopping mall in Nagoya CBD (one of the largest in Japan). It uses data from a gate count survey of pedestrian passage volume on weekday afternoons in which the underground streets are viewed as a unified closed space. In the underground shopping mall, there are ticket gates for two stations on three underground lines. Of the three factor models that are adopted, the first and second factors are the shortest distances to the station ticket gates and to the buildings connected underground. However, it is noteworthy that both are not physical distances but phase distances (Visual Depth Step). The third factor is the number of nearby shops; this point means that the visible should not be overlooked.

In Chap. 5, the Ota–Kaneda study is based on the drastically renewed street networks, block patterns, and buildings in central Nagoya after WWII; this article conducts an analysis of the causes of prosperity in the 1934 and 1964 eras. However, as it is impossible to obtain direct crowding data, land rent and route price data are used as alternative indicators. Examining the adopted factors, global integration value (GIV) has been found in both eras in addition to the agglomeration factor, infrastructure factor (street width), and accessibility factor (electric stop). Consequently, there is clearly a significance to the factors of shape.

The Ota–Takahashi–Kaneda paper in Chap. 6 uses REINS office rent data (which is a common database for real estate transactions) to analyze the factors in the central Tokyo Kanda Station 800 m zone where many relatively small offices are located. The adopted model includes variables of age, number of shops, contract size, and distance from stations, while the VGA index is not used. This shows that the VGA index is not an all-purpose index, and instead, that visibility is not necessarily a factor in the context of office rent.

Agent Modeling of the Downtown Visitor and Dynamic Simulation of the Downtown Area

The third question regards both the design principles of visitor agents who perform the shop-around behaviors and the downtown dynamics, which is a multi-agent model that represents the visitor agents. This is a continuation of both the first question and its answer. As mentioned above, the results of the data collection and analysis on visitors' shop-around behaviors were obtained. The content of this question concerns how the modeling and simulation for the downtown dynamics structure can be designed, in particular, the visitor agent model that simulates shop-around behavior. Here, it is worth stating that if downtown dynamics assume that both the agent population configuration and the interaction system are set unchanged, the driving factors of the downtown area ultimately originate only from external environmental conditions and updated internal information; therefore, it is necessary to consider these points in advance in the context of agent model design.

Dense grounding and intervention in agent modeling means abandoning the KISS principle and aiming for heavy (structural) agents, but this raises another question in this modeling study as to how to allow for bounded rationalities and intellectual functions in the structure of agents. Additionally, this approach has not yet been established (except in some advanced fields), and the formation of bold logic is required through free explorative ideas.

The model in this book is referred to as an errand-based approach that draws on the concept of activity-based models, which includes everyday travel behavior models. The first characteristic of this approach includes the visitor agents as decision-making subjects who plan and implement multiple activities (errands) when visiting commercial districts; this assumes a microcause in that time performance will affect the preferences for the next district visit. In visitor decision making, two novel models are adopted for the concept of errands. The first is a garbage-can model that determines departures when the variety of errands exceeds a threshold value using a model of visitor decision making: this differs from Poisson models with fixed departure ratios for time using a LLPM (Linked Logit and Poisson

Model). The other differs from existing spatial interaction models in that it replaces the commercial accumulation size with the number of shop types that allow errands to be accomplished using a model for selecting a visit destination from plural candidate commercial districts.

The second characteristic regarding visitors' spatial behaviors inside the downtown area is the assignment of shop visits within a time budget, i.e., the formation of a schedule plan and the improvised actions (opportunistic planning) inside the downtown area. In the results, the shop-around behavior is categorized into planned action and improvised action; these are further divided into alternative action, impulse visit, and detour walking. In particular, the planned action occurs in accordance with the schedule, whereas the alternative action to the improvised action is an alternative visit to another shop when the planned errand is not achieved. To that end, the model aims to introduce a dynamic scheduling function, but it exceeds the Markov model on which the existing shop-around behavior model research was based.

The third characteristic is the handling of dynamic updates inside agents using rescheduling, which is also related to intellectual functions in terms of adaptation and learning. This has a bounded rationality of not assuming knowledge, but it is characterized by dynamic updating in terms of rescheduling.

The Yoshida paper in Chap. 7 explains the composition of an agent-based simulation in connection with the ASSA project that was conducted between 2005 and 2010; the paper is a partial reprint of Mr. Takumi Yoshida's doctoral thesis, which describes the design up to ASSA ver. 3. The chapter explains the structure of a visitor agent model that characterizes MPMS behaviors in the downtown area as clarified in the Osu district data analysis mentioned above.

In Chap. 8, the Yoshida–Kaneda research demonstrates example simulations of visitor agent shop-around behavior in the Osu district using ASSA ver. 3 (as explained in Chap. 7) and data from the 2003 Osu district survey. It also reports the simulations for some policy scenarios (including tenant relocation) alongside a performance evaluation.

The Chap. 9, Kaneda–Shohmitsu–Chang study reports the modeling and simulation of downtown dynamics in an approach toward the notion of artificial society. This downtown dynamics model is a prototype that deals with systems of interaction between customer (visitor) agents to downtown districts and also shop (tenant) agents who conduct business activities. A downtown customer agent model is simpler than ASSA, and a shop tenant agent is introduced with the expected benefit maximization principle for selecting shop types for each shop opening. In the downtown dynamics model using simulations based on a virtual district example, the variety of shop types in the downtown area shows the possibility of increased sustainability to compete with large-scale suburban stores. Additionally, based on the simulation experiments that test Jacobs' conditions for city diversity, theoretical implications are obtained (e.g., the positive effect on sustainability as a result of the increase in city diversity according to smaller block sizes).

Applicability of Vision-Driven Agents

The fourth question expands on the second question and answer and asks: what in our agent models should be explored as the research initiative to base human microscopic behavior in urban space from the perspective of visibility?

This book presents the possibilities in both theory and practice for vision-driven pedestrian agents. It does so via a mechanism whereby pedestrian agents search and internally aggregate vision fields to create reference points for use in selecting actions. The vision-driven agent is not only promising for crowding simulation in the middle-scale urban space mentioned above, but it is also the mechanism that handles visual recognition (vision-driven mechanism), which itself becomes a source of bounded rationality in human spatial action selection; it is also indicative for the bounded rationality studies. Each of the vision-driven agents is a microscopic model, and its grounding is assumed to utilize experimental data or introspection based on intrinsic human experience in addition to survey data.

Natural movement is another space syntax theory that is focused on here. It is based on Hillier's discussion in the 1990s whereby humans (as pedestrians with no destination and no prior knowledge of the space) choose spatial behavior based only on a given spatial shape reflected in their field of vision. Turner and Penn's EVA (exosomatic visual architecture) model proposed in 2004 is accepted as the basic model. EVA can be interpreted as a model for expressing the free exploration of pedestrians with no destination. The primary reason to focus on this relates to research reports that reveal a high correlation between the spatial distribution of pedestrian agents obtained from EVA simulations with the distribution in actual middle-scale urban spaces. Penn and Turner reported a strong correlation with EVA simulation results in comparison with passage volume data at a department store in London; this book's research group also confirmed a strong correlation with EVA simulation results in comparison with encounter survey results recently conducted in the Kanayama district of Nagoya.

Furthermore, the hypothesis that natural movement can be a normative model of redundancy based on visibility is also considered. Walking behavior is often modeled by the shortest path, but in small- to middle-scale urban space it is normal that the actual walking path followed has redundancy with respect to the shortest path, which is known to be characterized by visibility. Although mathematical examinations await future development, the EVA algorithm is based on a different idea from the random walk (a mathematical concept of stochastic process theory), which is attractive as a candidate for use as a pure redundancy model. In the approach that considers a bounded rationality as a deviation from rationality, a model that combines an optimal model of the shortest route to the destination with this redundant model seems attractive when modeling the redundant walking behavior of a pedestrian with a destination.

In most of the existing agent-based simulation research (including the research previously discussed), the two-dimensional space has been used in which an agent's sight fields and behaviors are expressed. In the small- to middle-scale urban space it is easy to find elements that require a height dimension to deal with agents' behaviors; for this reason, the concept of artificial society is oriented to the extended implementation of the three-dimensional space. Although these kinds of pedestrian agent models are inspired by a somewhat unusual thinking experiment, they are expected to reveal an unexplored field of human dimension aspects of urban space.

The Susaki–Kaneda paper in Chap. 10 explains a redesigned version of VD-Walker as a vision-driven agent designed on Unity (a game engine that handles three-dimensional space) after reviewing the original EVA by Turner and Penn. In addition, VD-Walker can handle weighted OD by Karimi's group concerning navigational movements defined as spatial behaviors with given destinations and which include redundancy with vision-driven mechanisms.

In Chap. 11, the Kaneda–Morita–Ohashi–Mizuno paper reports VDPA1, which is a vision-driven agent engaged with prior to the VD-Walker mentioned in Chap. 10. This agent also deals with navigational movement via simulations with a focus on the trajectory pattern of pedestrian agents walking along curved paths to avoid furniture (such as chairs, tables, and plantings) in public spaces. Interestingly, this simulation forms a piece of furniture with few walk-throughs at the same time as the formation of the agents' walking trajectories. Both of these have a figure–ground relationship, and the model demonstrates the possibility of using VD-agent simulation in architectural planning.

The Kino–Ohashi–Kaneda study in Chap. 12 is a research report that built a 3D virtual space as a tool for collecting behavior data in VD-agent modeling and analyzed the behavior in the space through gaming experiments. In addition to the ease of collecting both action and behavioral log data in the system, the advantage of this experimental method is that cognitive and behavioral parameters can be collected from a relatively small sample. For example, augmenting the notion of framing for focused cognitive targets in advance integrates with the agent-based simulation in virtual space. This research attempts to provide a foundation of visibility for the concept of personal spacing via an example situation for selecting a waiting place in a public space.

In Chap. 13, the Yokoyama–Maekawa–Kaneda paper simulates pedestrians' wayfinding in a public space using sign layout setups; this is a promising field in the theoretical and practical application of vision-driven agent simulation. In a situation in the station square space expressed in 3D on Unity, the pedestrian agent performs navigational movement toward their own destination by sequentially following the visually recognized signs. In this paper, the results show that the bifurcation that drastically changes the route choice (as a result of wayfinding behavior) is caused by the minute difference in the initial position of the agent and the position of the sign layout.

In *Downtown Dynamics*, one perspective emphasizes breadth over depth; however, rather than being strictly research programs, they are also research initiatives in open communities that represent the diverse activities that take place in downtown areas. Therefore, *Downtown Dynamics* is an approach not only for researchers but also for citizens based on citizen values that facilitates grounding and policy interventions depending on the realities that everybody experiences. In that sense, this book's aim is citizen sciences. The editor is reliant on the reader's expectation for the research of free explorative ideas based on the notions of feeling and inspiration, which are founded in interaction.

Graduate School of Engineering
Nagoya Institute of Technology
Nagoya, Aichi, Japan

Toshiyuki Kaneda

Acknowledgments

This book was originally planned as a monograph but as a result it became a collection of research papers written mainly between 2010 and 2019. Each chapter can be read independently. Many of the chapters were written by students/graduates who collaborated on research projects at the authors' laboratory and with the editor. Also, two authors were invited from outside the laboratory to participate in writing this book. These are Professor Masakazu Takahashi and Dr. Masaki Kitazawa (Chap. 3), Professor Hiroshi Takahashi (based on a collaborative research with Dr. Akira Ota) (Chap. 6), and Professor Kasumi Susaki (based on a collaborative research with Kaneda) (Chap. 10). Our thanks to Ms. Sayori Harada for editing the publication. Part of the publishing was supported by JSPS KAKENHI, grant number 18H03825.

Contents

About the Editor

Toshiyuki Kaneda is a Professor at the Graduate School of Engineering, Nagoya Institute of Technology. Holding a Doctor of Engineering from Tokyo Institute of Technology, he has served as a special researcher for the JSPS (Japan Society for the Promotion of Science) and an Assistant Professor at Tokyo Institute of Technology. A fellow and former chairman of JASAG (Japan Association of Simulation And Gaming), his current research interests include Agent-Based Urban Simulation and Urban Analytics.

Part I
Downtown as Phenomenon and Its Mechanism

Chapter 1
A Review of a Shop-Around Behavior Survey in the Osu District

Toshiyuki Kaneda and Sohei Inagaki

Abstract In this chapter, the framework of a shop-around behavior survey conducted in the Osu district is explained. A shop-type configuration survey conducted together with the basic survey and an advanced analysis of these data are also discussed. In addition, the results of the past four surveys conducted by the Kaneda laboratory are reviewed, and the visiting behavior of the district's visitors is shown at each time point along with their spatial distributions of walkthrough and visit frequencies. Furthermore, the shop-type configuration surveys (each of which is indispensable for analyzing the interactions between visitor behavior and shops) are also assessed. The survey results at the four points are expected to analyze the dynamics in the Osu district in the form of panel data. Additionally, the 2018 survey categorized the customer base of visitors to the district into six clusters and confirmed the existence of multiple customer bases with different characteristics.

1.1 Introduction

Many commercial districts in Japan have fallen into decline over the past 50 years. Among the numerous commercial areas dominated by small scale shops within the Osu district in Nagoya, there are lively commercial districts characterized by an agglomeration and specialization of shop types. The crowds of shoppers drawn by these stores have been maintained because diverse visitor groups have been attracted; however, the configuration of the shops in these districts has continued to change—depending on the time period. The Osu district is a pool of diversity for

T. Kaneda (✉)
Graduate School of Engineering, Nagoya Institute of Technology, Gokiso, Showa, Nagoya, Japan
e-mail: kaneda@nitech.ac.jp

S. Inagaki
Department of Architecture and Design, Nagoya City Office, Sannomaru, Naka, Nagoya, Japan
e-mail: s.inagaki.12@city.nagoya.lg.jp

T. Kaneda (ed.), *Downtown Dynamics*, Agent-Based Social Systems 16,
https://doi.org/10.1007/978-4-431-54901-7_1

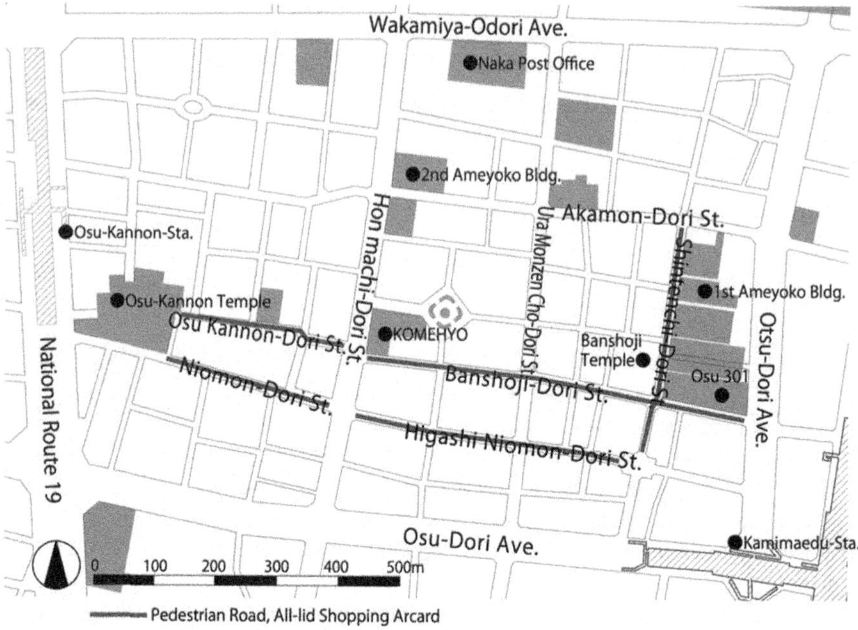

Fig. 1.1 Target Survey Area: Osu district, Nagoya, Japan

both shop types and visitor groups. Such pools are sources of adaptability to temporal and environmental changes.

The shop-around behavior of visitors walking between shops (or, more generally, between facilities) in the district refers to a unit of micro-behavior that expresses the interaction between individual visitors and individual shops; the whole district can be observed by agglomerating these units as mesoscale phenomena. The attraction of analyzing data from district visitors and shop-type configurations lies in the fact that the mechanisms of crowds can be clarified as urban mesoscale phenomena, and other findings can be obtained to study the stimulation of downtown areas. Therefore, the concept of shop-around behavior is key to unraveling the mechanisms of crowds in such districts. In light of the above, the Kaneda laboratory at the Nagoya Institute of Technology conducted questionnaire surveys regarding shop-around behavior among visitors to the Osu district in 1998, 2003, 2008, 2013, and 2018.

The Osu district (Fig. 1.1) was the target of the Kaneda laboratory survey and is an interconnected downtown area south of Sakae, Nayabashi, and Fushimi in Nagoya and is enclosed by Wakamiya-Odori Ave. Originally, this area was built around a temple/shrine along Honmachi Dori St. during the Edo period. Osu Kannon is located on the west side of this district, and there are several popular and crowded shopping streets set out in a grid-like arrangement.

The survey target was the agglomeration of more than 500 retail/food and drink shops within the area enclosed by Otsu Dori Ave., Osu Dori Ave., National Route 19 (Fushimi Dori St.), and Wakamiya-Odori Ave. (approx. 30 ha). Within this district, a particular focus was placed on the nine key streets and their many shops that form the backbone of this area (Akamon Dori St., Osu Kannon Dori St., Banshoji Dori St., Niomon Dori St., Higashi Niomon Dori St., Honmachi Dori St., Uramonzencho Dori St., Kita-Shintenchi Dori St., and Shintenchi Dori St.) along with Otsu Dori Ave. (west side), which is the main street adjacent to the Osu district.

In addition to explaining the outline of the shop-around behavior survey conducted in Osu, this section discusses the data from the shop-type configuration survey that was also conducted as well as the basic statistics and analyses of this data. In addition, the study provides a summary of the results of the last four surveys conducted by the Kaneda laboratory. In terms of the survey results from 2018 in particular, the results of the district visitor-cluster analysis are discussed.

1.2 Explanation of the Survey on Shop-Around Behavior in the Osu District

1.2.1 Shop-Around Behavior Survey Method

Shop-around behavior refers to visitors walking around (intra-district trips) between shops (or, more generally, between facilities) in a district. Well-known studies of shop-around behavior include those by Timmermans' group in Eindhoven in the 1980s and the studies by Saito's group in Fukuoka (Timmermans 2005, 2009; Saito 1984, 1998). The nature of a multipurpose multi-stop (MPMS) event, which signifies the planning and adaptability of people's behavior, becomes even clearer in comparison with the person–trip survey, which deals with daily transport actions in urban areas. Shop-around behavior data refers to data on connected trips within a district.

The methods for collecting data on walking around between shops in a district include (1) interviews, (2) questionnaires, (3) visual observations, (4) digital sensing, and (5) social media. The advantage of questionnaire surveys is that data can be gathered on a visitor's planned and unplanned actions, which is difficult with visual observation and digital sensing methods. Furthermore, questionnaire surveys have no problems in terms of consent to individual behavior data compared with visual observation and digital sensing, and questionnaires are advantageous over interview methods, which are more labor-intensive. Questionnaire surveys are likely to remain an effective method until the full-scale application of social media.

A questionnaire was used for the shop-around behavior survey. Questionnaires were distributed among visitors to the district during daytime on autumn weekends. Responses were collected by mail after respondents completed the questionnaire at home (on-site distribution/mail collection). Respondents were questioned about visitor attributes, shop visits, visit plans, and more, and were asked to plot their walking route on a map of the district. Answers relating to attributes included gender, age, number/type of companions, means of transport, and so on. Respondents used a map to show both their walking route and the location/order of shops visited. The questions on the shops visited included the shop type, the duration of visit, whether the visit was planned in advance (before coming to the district), and whether products and/or services were purchased. The content of the questionnaire is shown in Fig. 1.2. This method allowed for the statistical analysis of planned and unplanned visitor actions, which is difficult to do with follow-up surveys and on-site interviews. In the 2018 survey, 38 responses were collected using a web survey with the same content.

At the Kaneda laboratory, in addition to conducting multi-point distribution surveys in the Osu district in 1998, 2003, 2008, 2013, and 2018, results of multi-point distribution district visitor surveys were obtained from the Nagoya Station (east) district (Misaka et al. 2001), Hirokoji (Misaka and Kaneda 2005), all of central Nagoya (Arakawa and Kaneda 2002), Akihabara (Hayashi et al. 2013), and more. Additionally, results from single-point distribution facility visitor surveys were included from (among others) the former Asunal Kanayama (Misaka et al. 2010), the music theater (Misaka and Kaneda 2003), the traditional theater (Kitamura et al. 2003), the Nagoya City Art Museum (Misaka and Kaneda 2006), and Denki Bunka Kaikan (cultural facilities).

1.2.2 Summary of Each Survey and Basic Statistics Regarding Respondents' Data

Shop-around behavior survey data from respondents was first aggregated by the number of respondents per external attribute. Here, the external attributes refer to gender, age, and residence on the "Face Sheet," as well as the number of companions, frequency of visits to the district, and the means of transport. Next, the duration of visit, walking distance, and the number of shops visited (the basic characteristic variables of shop-around behavior) were aggregated by external attributes (Table 1.1).

The visit count per visitor for each shop type was calculated. Table 1.2 contains the ranking of visit counts per visitor. In the Osu district, the shops visited in each survey varied greatly reflecting the specific demands in each area within Osu. Below, in addition to summarizing each of the Osu district surveys, the characteristics of personal attributes of survey respondents are described.

1. **1998 survey**: In the initial survey, 2250 questionnaires were distributed; 184 were collected, giving a collection rate of 8.18%. It is worth noting that a greater ratio of respondents were men (64:36 men to women) and electronics shops were the most visited shop type (2.11). Significant behavioral characteristics included the fact that the main cluster of respondents in this survey was identified as the "male nerd cluster," while the shop-around corridor pattern (described later) was loop-A between Akamon Dori St.–Honmachi Dori St.–Banshoji Dori St., and Shintenchi Dori St.

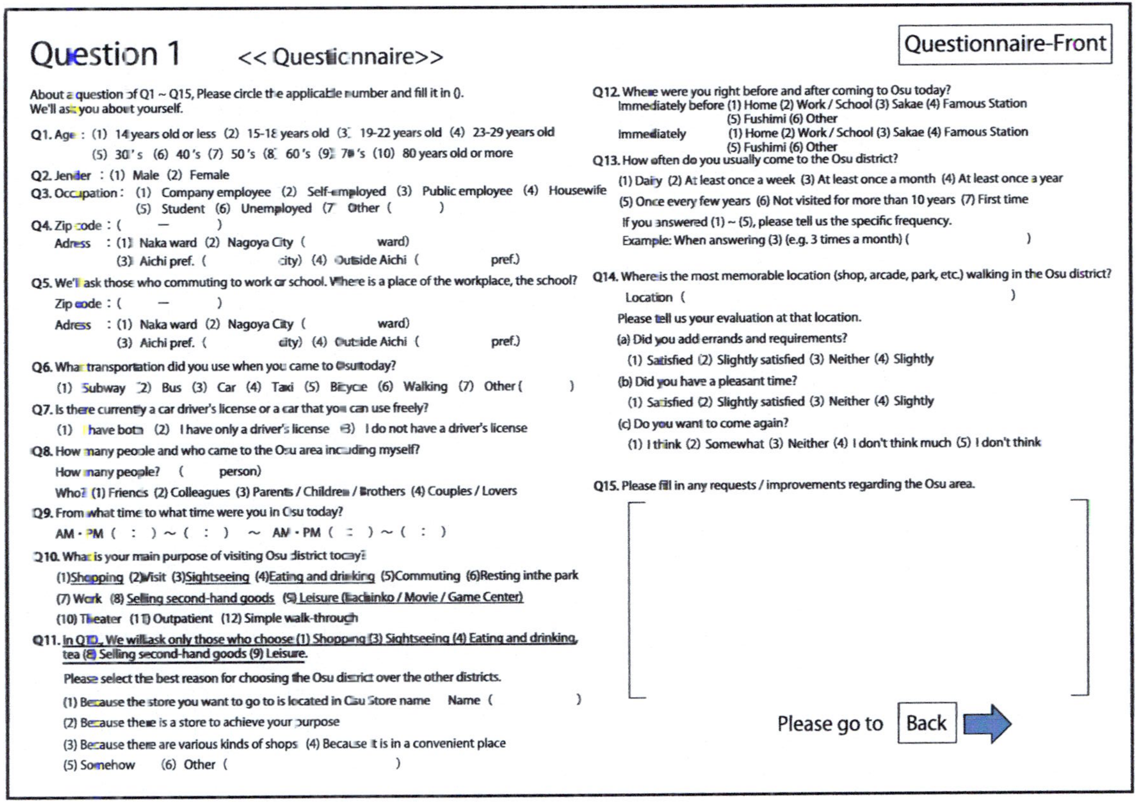
Question 1 << Questionnaire>>

Questionnaire-Front

About a question of Q1 ~ Q15, Please circle the applicable number and fill it in ().
We'll ask you about yourself.

Q1. Age : (1) 14 years old or less (2) 15-18 years old (3) 19-22 years old (4) 23-29 years old
(5) 30's (6) 40's (7) 50's (8) 60's (9) 70's (10) 80 years old or more

Q2. Jender : (1) Male (2) Female

Q3. Occupation : (1) Company employee (2) Self-employed (3) Public employee (4) Housewife
(5) Student (6) Unemployed (7) Other ( )

Q4. Zip code : (—)
Adress : (1) Naka ward (2) Nagoya City (ward)
(3) Aichi pref. (city) (4) Outside Aichi (pref.)

Q5. We'll ask those who commuting to work or school. Where is a place of the workplace, the school?
Zip code : (—)
Adress : (1) Naka ward (2) Nagoya City (ward)
(3) Aichi pref. (city) (4) Outside Aichi (pref.)

Q6. What transportation did you use when you came to Osu today?
(1) Subway (2) Bus (3) Car (4) Taxi (5) Bicycle (6) Walking (7) Other ( )

Q7. Is there currently a car driver's license or a car that you can use freely?
(1) I have both (2) I have only a driver's license (3) I do not have a driver's license

Q8. How many people and who came to the Osu area including myself?
How many people? (person)
Who? (1) Friends (2) Colleagues (3) Parents / Children / Brothers (4) Couples / Lovers

Q9. From what time to what time were you in Osu today?
AM・PM (:) ~ (:) ~ AM・PM (:) ~ (:)

Q10. What is your main purpose of visiting Osu district today?
(1)Shopping (2)Visit (3)Sightseeing (4)Eating and drinking (5)Commuting (6)Resting inthe park
(7) Work (8) Selling second-hand goods (9) Leisure (Pachinko / Movie / Game Center)
(10) Theater (11) Outpatient (12) Simple walk-through

Q11. In Q10., We will ask only those who choose (1) Shopping (3) Sightseeing (4) Eating and drinking, tea (8) Selling second-hand goods (9) Leisure.
Please select the best reason for choosing the Osu district over the other districts.
(1) Because the store you want to go to is located in Osu Store name Name ( )
(2) Because there is a store to achieve your purpose
(3) Because there are various kinds of shops (4) Because it is in a convenient place
(5) Somehow (6) Other ( )

Q12. Where were you right before and after coming to Osu today?
Immediately before (1) Home (2) Work / School (3) Sakae (4) Famous Station
(5) Fushimi (6) Other
Immediately (1) Home (2) Work / School (3) Sakae (4) Famous Station
(5) Fushimi (6) Other

Q13. How often do you usually come to the Osu district?
(1) Daily (2) At least once a week (3) At least once a month (4) At least once a year
(5) Once every few years (6) Not visited for more than 10 years (7) First time
If you answered (1) ~ (5), please tell us the specific frequency.
Example: When answering (3) (e.g. 3 times a month) ( )

Q14. Where is the most memorable location (shop, arcade, park, etc.) walking in the Osu district?
Location ( )
Please tell us your evaluation at that location.
(a) Did you add errands and requirements?
(1) Satisfied (2) Slightly satisfied (3) Neither (4) Slightly
(b) Did you have a pleasant time?
(1) Satisfied (2) Slightly satisfied (3) Neither (4) Slightly
(c) Do you want to come again?
(1) I think (2) Somewhat (3) Neither (4) I don't think much (5) I don't think

Q15. Please fill in any requests / improvements regarding the Osu area.

Please go to Back

Fig. 1.2 Osu shop-around survey questionnaires

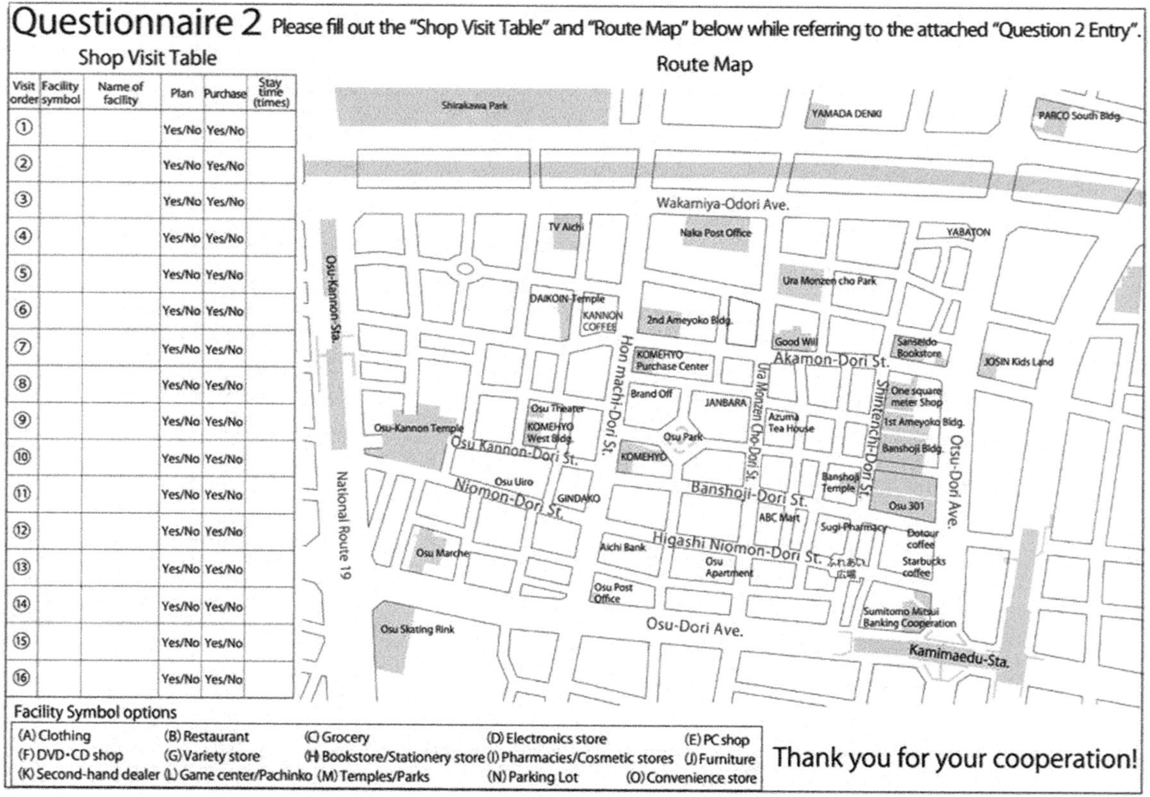

Questionnaire 2 Please fill out the "Shop Visit Table" and "Route Map" below while referring to the attached "Question 2 Entry".

Shop Visit Table

Visit order	Facility symbol	Name of facility	Plan	Purchase	Stay time (times)
①			Yes/No	Yes/No	
②			Yes/No	Yes/No	
③			Yes/No	Yes/No	
④			Yes/No	Yes/No	
⑤			Yes/No	Yes/No	
⑥			Yes/No	Yes/No	
⑦			Yes/No	Yes/No	
⑧			Yes/No	Yes/No	
⑨			Yes/No	Yes/No	
⑩			Yes/No	Yes/No	
⑪			Yes/No	Yes/No	
⑫			Yes/No	Yes/No	
⑬			Yes/No	Yes/No	
⑭			Yes/No	Yes/No	
⑮			Yes/No	Yes/No	
⑯			Yes/No	Yes/No	

Facility Symbol options

(A) Clothing (B) Restaurant (C) Grocery (D) Electronics store (E) PC shop
(F) DVD・CD shop (G) Variety store (H) Bookstore/Stationery store (I) Pharmacies/Cosmetic stores (J) Furniture
(K) Second-hand dealer (L) Game center/Pachinko (M) Temples/Parks (N) Parking Lot (O) Convenience store

Route Map

Thank you for your cooperation!

Fig. 1.2 (continued)

Table 1.1 Composition of respondents' attributes and basic characteristics of their behaviors

		Number (persons)	Rate (%)	Average stay time (min)	Average walking distance (km)	Average visited facilities (times)
1998	Whole	153	100%	140	1.34	5.40
Gender	Male	96	64%	143	1.34	5.60
	Female	55	36%	173*	1.33	5.50
Age	~29	46	30%	167	1.41	6.40
	30–49	58	38%	153	1.38	5.90
	50~	49	42%	150	1.22	4.30**
2003	Whole	491	100%	158	1.14	4.98
Gender	Male	231	48%	168	1.17	5.07
	Female	253	52%	142	1.12	4.89
Age	~29	120	25%	177	1.13	5.12
	30–49	215	44%	155	1.18	5.33
	50~	152	31%	150	1.09	4.29
2008	Whole	607	100%	167	1.46	4.47
Gender	Male	249	42%	162	1.40	4.27
	Female	349	58%	170	1.51	4.63
Age	~29	120	20%	188	1.54	4.61
	30–49	263	43%	172**	1.48	4.81**
	50~	223	37%	148	1.39	3.98
2013	Whole	**295**	100%	182	1.49	3.82
Gender	Male	155	53%	174	1.40*	3.59
	Female	140	47%	190	1.57	4.04
Age	~29	60	20%	163	1.39	3.75
	30–49	111	38%	184	1.52	4.21
	50~	124	42%	189	1.50	3.50
2018	Whole	505	100%	158	1.48	3.70
Gender	Male	257	53%	152	1.43	3.57
	Female	231	47%	162	1.54	3.91
Age	~29	55	11%	203	1.48	3.91
	30–49	152	31%	172**	1.66**	4.11
	50~	288	58%	140	1.36	3.41

(mean Difference Test: **Significant difference 1%; *significant difference 5%)

2. **2003 survey**: In the second survey, 3000 questionnaires were distributed; 504 were collected making this a full-scale survey with a collection rate of 16.8%. Following this survey, a distribution plan (sites and number of distributions) was established to conduct a walkthrough volume survey prior to multi-point distribution. The male-to-female ratio of respondents was 48:52, with shop-type visits by women being characterized by clothing shops (0.93) and food and drink shops (0.86), in that order. Furthermore, based on the location of the large car park between Otsu Dori Ave. and Shintenchi Dori St. and the major increase in clothing shops (138–>191) and food and drink shops (105–>126), the prominent cluster of respondents was the "female cluster making multiple visits by car." The corridor pattern, which is described later, was primarily loop-B

Table 1.2 Visited shop type per visitor in the five surveys (numbers in the table are provisional)

		Year	First place		2nd place		3rd place		4th place		5th place		6th place		7th place		8th place	
Whole		1998	Electronics	2.11	Clothing	0.59	Restaurant	0.59	Rec., Complex	0.49	Var., Furn.	0.44	Temple, park	0.34	Grocery	0.27	Others	0.25
		2003	Electronics	1.05	Restaurant	0.84	Clothing	0.64	Rec., Complex	0.57	Var., Furn.	0.53	Temple, park	0.33	Others	0.25	Grocery	0.24
		2008	Restaurant	0.94	Clothing	0.85	Rec., Complex	0.65	Var., Furn.	0.64	Temple, park	0.45	Grocery	0.42	Electronics	0.30	Others	0.21
		2013	Restaurant	0.92	Var., Furn.	0.82	Clothing	0.63	Electronics	0.61	Others	0.52	Temple, park	0.42	Grocery	0.29		
		2018	Restaurant	1.00	Others	0.55	Var., Furn.	0.55	Grocery	0.47	Temple, park	0.42	Clothing	0.40	Electronics	0.28		
Gender	Male	1998	Electronics	2.79	Restaurant	0.48	Rec., Complex	0.44	Var., Furn.	0.39	Clothing	0.28	Temple, park	0.27	Others	0.27	Grocery	0.17
		2003	Electronics	1.78	Restaurant	0.80	Rec., Complex	0.50	Var., Furn.	0.46	Clothing	0.33	Others	0.28	Temple, park	0.28	Grocery	0.20
		2008	Rec., Complex	0.84	Restaurant	0.79	Var., Furn.	0.63	Electronics	0.54	Clothing	0.47	Temple, park	0.47	Grocery	0.28	Others	0.25
		2013	Electronics	0.95	Restaurant	0.65	Var., Furn.	0.60	Others	0.46	Clothing	0.40	Temple, park	0.38	Grocery	0.19		
		2018	Restaurant	0.82	Others	0.61	Var., Furn.	0.60	Grocery	0.43	Electronics	0.42	Temple, park	0.36	Clothing	0.30		
	Female	1998	Clothing	1.14	Electronics	0.89	Restaurant	0.79	Rec., Complex	0.58	Var., Furn.	0.53	Temple, park	0.46	Grocery	0.46	Others	0.25
		2003	Clothing	0.93	Restaurant	0.86	Rec., Complex	0.63	Var., Furn.	0.59	Temple, park	0.39	Electronics	0.38	Others	0.33	Grocery	0.27
		2008	Clothing	0.13	Restaurant	1.05	Var., Furn.	0.64	Grocery	0.53	Rec., Complex	0.50	Temple, park	0.42	Others	0.19	Electronics	0.13
		2013	Restaurant	1.12	Var., Furn.	0.99	Clothing	0.88	Others	0.54	Temple, park	0.42	Grocery	0.37	Electronics	1.14		
		2018	Restaurant	1.21	Others	0.51	Grocery	0.51	Var., Furn.	0.50	Temple, park	0.49	Clothing	0.49	Electronics	1.14		

Age	~29	1998	Electronics	2.83	Clothing	0.98	Var., Furn.	0.76	Rec., Complex	0.46	Restaurant	0.43	Others	0.28	Temple, park	0.26	Grocery	0.15
		2003	Clothing	1.13	Electronics	1.10	Restaurant	0.89	Var., Furn.	0.62	Rec., Complex	0.41	Temple, park	0.28	Others	0.25	Grocery	0.15
		2008	Clothing	1.29	Restaurant	0.93	Var., Furn.	0.76	Rec., Complex	0.59	Electronics	0.35	Others	0.26	Temple, park	0.23	Grocery	0.21
		2013	Var., Furn.	1.00	Clothing	0.85	Restaurant	0.82	Others	0.41	Electronics	0.39	Temple, park	0.16	Grocery	0.08		
		2018	Restaurant	1.29	Var., Furn.	0.86	Others	0.66	Clothing	0.51	Electronics	0.40	Temple, park	0.14	Grocery	0.03		
	30–49	1998	Electronics	2.53	Rec., Complex	0.63	Restaurant	0.58	Clothing	0.35	Var., Furn.	0.33	Temple, park	0.28	Grocery	0.22	Others	0.20
		2003	Electronics	1.23	Restaurant	0.83	Rec., Complex	0.63	Var., Furn.	0.53	Clothing	0.52	Temple, park	0.29	Grocery	0.28	Others	0.22
		2008	Restaurant	1.10	Clothing	0.92	Var., Furn.	0.77	Rec., Complex	0.67	Grocery	0.39	Temple, park	0.38	Electronics	0.32	Others	0.24
		2013	Restaurant	1.12	Var., Furn.	0.94	Clothing	0.73	Electronics	0.60	Others	0.55	Temple, park	0.33	Grocery	0.22		
		2018	Restaurant	1.15	Others	0.74	Var., Furn.	0.61	Grocery	0.50	Clothing	0.45	Temple, park	0.40	Electronics	0.29		
	50~	1998	Electronics	0.96	Restaurant	0.74	Clothing	0.56	Temple, park	0.48	Grocery	0.43	Rec., Complex	0.37	Others	0.28	Var., Furn.	0.24
		2003	Electronics	0.73	Restaurant	0.72	Rec., Complex	0.60	Var., Furn.	0.46	Temple, park	0.43	Clothing	0.43	Others	0.29	Grocery	0.24
		2008	Restaurant	0.76	Rec., Complex	0.66	Temple, park	0.64	Grocery	0.59	Clothing	0.52	Var., Furn.	0.40	Electronics	0.26	Others	0.15
		2013	Restaurant	0.78	Electronics	0.62	Temple, park	0.62	Var., Furn.	0.61	Others	0.56	Grocery	0.46	Clothing	0.43		
		2018	Restaurant	0.85	Grocery	0.55	Temple, park	0.49	Var., Furn.	0.45	Others	0.41	Clothing	0.34	Electronics	0.25		

between Banshoji Dori St.–Osu Kannon Dori St.–Niomon Dori St., and Higashi Niomon Dori St., which continued to be the case thereafter.

3. **2008 survey**: In the third survey, 2079 questionnaires were distributed; 615 were collected giving a high collection rate of 29.7%. Following this survey, a collection rate of 20% or above has been maintained. One of the reasons for this is that the first survey was reported in a newspaper article the following day. Based on the improvements to the arcade on Higashi Niomon Dori St. and the successive positioning of large-scale second-hand goods shops on Osu Kannon Dori St. and Banshoji Dori St., there was a higher ratio of women than men among the respondents (male–female ratio: 42:58). The female cluster was the most prevalent cluster of respondents, with the overall visited shop types being food and drink shops (0.94), clothing shops (0.85), and second-hand goods shops (0.65).
4. **2013 survey**: In the fourth survey, 1346 questionnaires were distributed; 316 were collected, giving a collection rate of 23.5%. In terms of shop-type configuration, there was an increase in the number of variety and furniture shops (101–>127), and food and drink shops (130–>151). The number of clothing shops significantly decreased (183–>164 shops). The order of overall shop types visited was food and drink (0.92), variety and furniture shops (0.82), and clothing shops (0.63), and there was a transition in the preference of the female cluster from "boom in second-hand clothes" to "walking and eating on the street." There was also a record high in terms of the average stay time of 182 min (1998: 153 min, 2003: 148 min, 2008: 167 min).
5. **2018 survey**: In the fifth survey, 2000 questionnaires were distributed; 506 were collected giving a collection rate of 25.3%. Of the respondents, 60% were aged 60 or above, the average stay time decreased to 158 min, and the average number of shops visited reached a record low of 3.7 (1998: 5.4, 2003: 5.0, 2008: 4.5, 2013: 4.2) showing the clear presence of a new middle-aged and elderly local resident cluster.

A survey collection rate of 20% or more has been maintained since the 2008 survey. One of the reasons for this is improved awareness due to the accumulation of newspaper reports after each survey; however, the distribution of gifts (novelties) based on the survey results and the knowledge gained from the establishment of a distribution plan based on a prior walkthrough volume survey were also essential.

Importantly, with regard to survey bias, it should be noted that a gap appeared between the on-site reality of the composition of visitor age range and the age range composition of survey respondents in the fourth survey and thereafter. In terms of gender ratio, there was no discrepancy between the questionnaire and reality. Also, as the acceptance of questionnaires is influenced by minute factors, some inconsistencies were found in the actual distribution figures in connection with the distribution plan.

According to Table 1.2, with regard to overall visitors, the 1998 visit count per visitor was (in order) electronics shops (2.11), clothing shops (0.59), and food and drink shops (0.59); however, while the electronics shop-visit count decreased, the food-and-drink-shop-visit count increased, which led to a major change in 2018 to food and drink (1.00), leisure (0.55), and variety and furniture shops (0.55).

1.2.3 Shop-Type Configuration Survey and Shop-Visit Count

Shop-type configuration surveys were conducted incidentally to the shop-around behavior survey. The configuration surveys were based on an on-site survey (Fig. 1.3) and clarified the spatial distribution of first-floor shop types on the nine main streets. Until the 2003 survey, the study area covered the nine main streets based on the status of shop positioning along the roads in the district. From the 2008 survey onwards, the survey has also been conducted along the west side of Otsu Dori Ave.; a survey of the first-floor shop types along these 10 streets was conducted, and the shop types were categorized.

In the 2018 survey, shop types were categorized in eight shop-type groupings: food and drink, groceries, variety goods, furniture, electronics shops, clothing shops, temples/parks (for convenience, temples and parks are included among shops), and other.

The approach used in the shop coding system until the 2008 survey is different from the one in the 2013 survey and thereafter (at that time, the new codes were applied to the 2008 survey results for reference). There was also a change to the handling of second-hand goods shops/complex shopping facilities, which were categorized as second-hand goods/complex facilities shop types rather than the other category that had been used until 2008. However, this group was not suited to the changing times due to factors such as the addition of shops that were difficult

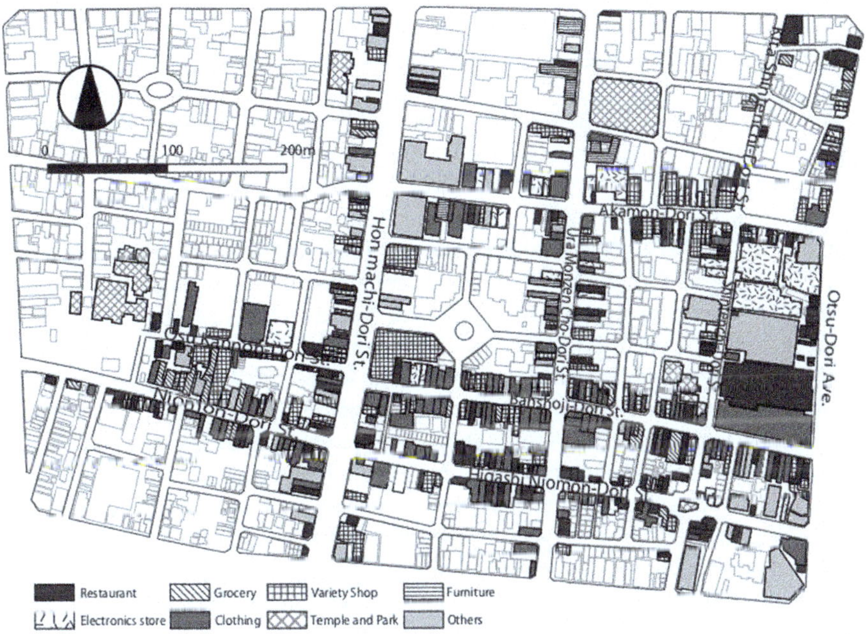

Fig. 1.3 Spatial distribution of first-floor shops along the main streets in Osu district in 2018

Table 1.3 Transition in the number of shops in the main shop types in each survey

		Restaurant and grocery	Variety shop and furniture	Clothing
1998	Number (shop)	(105)	(139)	(138)
	Fluctuation			
2003	Number (shop)	(126)	(91)	(191)
	Fluctuation	(Δ21)	(▼48)	(Δ53)
2008	Number (shop)	135 (130)	115 (101)	181 (183)
	Fluctuation	Δ9 (Δ4)	Δ24 (Δ10)	▼10 (▼8)
2013	Number (shop)	151	127	164
	Fluctuation	Δ16	Δ12	▼17
2018	Number (shop)	173	125	167
	Fluctuation	Δ22	▼2	Δ3

( ) are based on the 1998–2008 survey's count method

to categorize and the increasing diversification of tenants in complex facilities; therefore, in the 2013 survey and thereafter, the second-hand goods/complex facilities category was changed to other, and the second-hand goods shops were categorized depending on the type of products handled. For both the new and old codes, complex facilities are aggregated for one tenant on each floor with the shop type comprising the largest area on each floor calculated as the shop type for that floor. In addition, tenants on the first floor of shopping complexes with direct street access, opening hours that are not defined by the management of the shopping complex, and with separate tenant entrances (shutters) are counted as shops separate from the shopping complex.

The three main shop types in the Osu District are food and drink/groceries, clothing, and variety goods and furniture. In 1998, the order of these shops was variety goods and furniture (139 shops; 25.9%), clothing (138 shops; 25.7%), and food and drink/groceries (105 shops; 19.6%). Twenty years later, in 2018, this had changed to food and drink/groceries (173 shops; 30.7%), clothing (167 shops; 29.6%), and variety goods and furniture (115 shops; 20.4%). As clarified by a comparison between the changes in the configuration of visited shop types (which is described later), these transitions show flexible changes in the type of shops as driven by the changing demands over time (Table 1.3).

The main characteristics of the Osu district include the specialization and agglomeration of characteristic shop types on individual streets in a grid-like configuration. Table 1.4 reveals the shop-type configurations for each of the commercial districts. The results of the 2018 survey reveal an agglomeration of electronics stores on Akamon Dori St., clothing shops on Banshoji Dori St., variety goods/furniture shops on Honmachi Dori St. and Uramonzencho Dori St., and food and drink/grocery shops on Osu Kannon Dori St., Higashi Niomon Dori St. and Kita-Shintenchi Dori

Table 1.4 Transition in shop-type configuration on each street

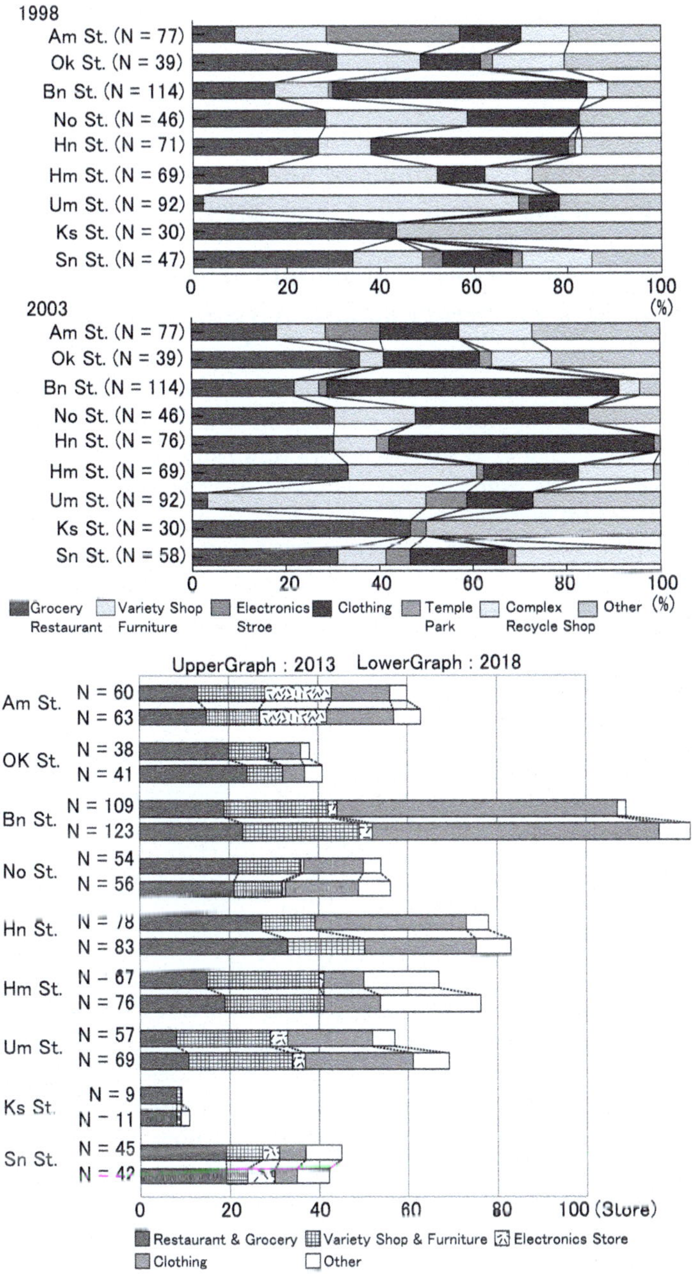

St.; however, the shifting trends over the last 5 years show that suitable changes have occurred, which demonstrates the complexity of the dynamism taking place in the background.

1.3 Features of Spatial Distributions of Shop-Around Behaviors in the 2003, 2008, 2013, and 2018 Surveys

1.3.1 Calculation of Spatial Distribution in the District for Walkthrough Frequency and Shop-Visit Frequency

Next, the shop-around behaviors and spatial distributions within Osu district are visualized using visitors' walking-route data and shop-visit data. Here, the main streets in the district are referred to as street segments in which roughly two–three blocks are divided into 27 units, with the total number of shops and the number of visitors shown for each street segment.

The definitions of the walkthrough frequency (visits/person) and shop-visit frequency (visits/person) are shown below.

- Average walkthrough frequency (visits/person) = sum total of the number of walking routes in each segment/number of samples
- Average shop-visit frequency (visits/person) = sum total of the number of visits to shops located in each segment/number of samples

Please note that the survey area covered the nine main streets until 2003, but the west side of Otsu Dori St. was added to the 2008 survey, making a total of 10 streets.

1.3.2 The 2003 Survey

Figure 1.4 contains the visitor-walkthrough frequency and shop-visit frequency data from the 2003 survey. The walkthrough frequency was highest in Osu Kannon Dori St. (2.78), Bansho-ji (2) (2.09), and Shintenchi Dori St. (1) (2.02), respectively. The shop-visit frequency was highest in Shintenchi Dori St. (1) (0.88), Banshoji (1) (0.71), and Osu Kannon (0.50), respectively. When comparing the frequency of shop visits with walkthrough frequency, a standout segment was Banshoji Dori St. (1). This street is home to large-scale stores (second-hand goods shops) that are scarce in the Osu district and thus attract large numbers of customers. In terms of the low frequency of shop visits compared with walkthrough frequency, the standout pass-through segments were Honmachi Dori St. (2), Bansho-ji (2), and Higashi Niomon Dori St. (2). The location of these streets is such that visitors pass through them on their way to their next target district (Oiwa et al. 2005).

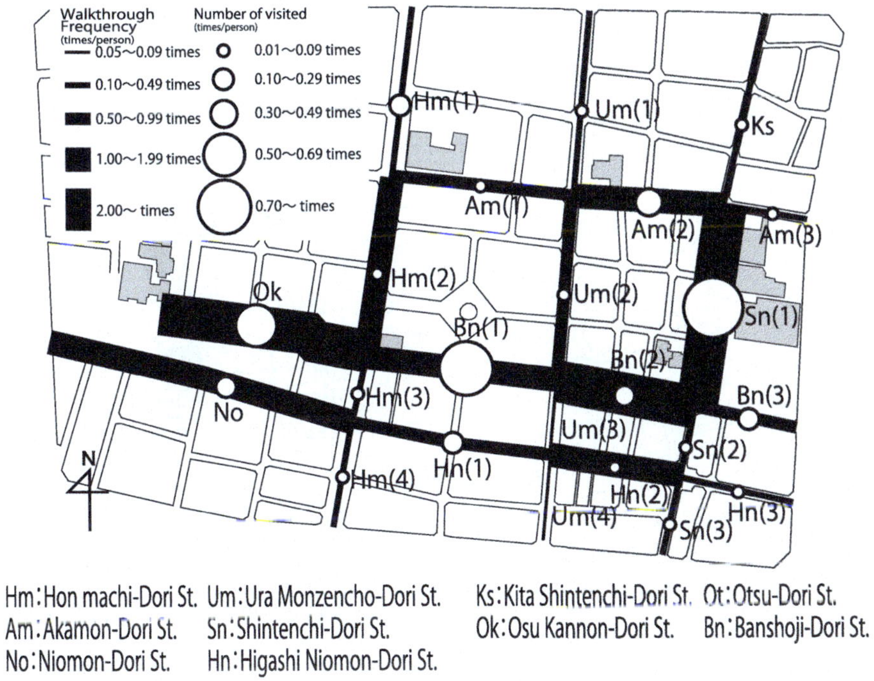

Fig. 1.4 Spatial distributions of visitors' walkthrough frequency and visit frequency in Osu in 2003 (Oiwa et al. 2005)

1.3.3 The 2008 Survey

Figure 1.5 shows the 2008 survey results. Walkthrough frequency was highest in Banshoji Dori St. (2) (0.74), Banshoji Dori St. (1) (0.71), and Osu Kannon Dori St. (0.62), respectively. The shop-visit frequency was highest in Osu Kannon Dori St. (0.65), Shintenchi (1) (0.50), and Bansho-ji (1) (0.43), respectively. The attractive segment with the highest frequency of shop visits compared with walkthrough frequency was Osu Kannon Dori St., which may be due to its many food and drink/grocery shops and the establishment of large second-hand goods shops. In addition, the significant pass-through segments for the low frequency of shop visits compared with walkthrough frequency were Honmachi Dori St. (2) and (3), and Shintenchi Dori St. (2).

When examining the 5-year changes after 2003, walkthrough frequency significantly increased in Higashi Niomon Dori St. (1) (0.35–>0.60) where repairs were made to the arcade. There were also increases in Banshoji Dori St. (3), Higashi Niomon Dori St. (2) and (3), and Niomon Dori St. However, there were decreases in Shintenchi Dori St. (1), Akamon Dori St. (1) and (2), and Banshoji Dori St. (1).

There was a significant increase in shop-visit frequency in Osu Kannon Dori St. (0.50–>0.65), yet there were major decreases in Banshoji Dori St. (1) (0.71–>

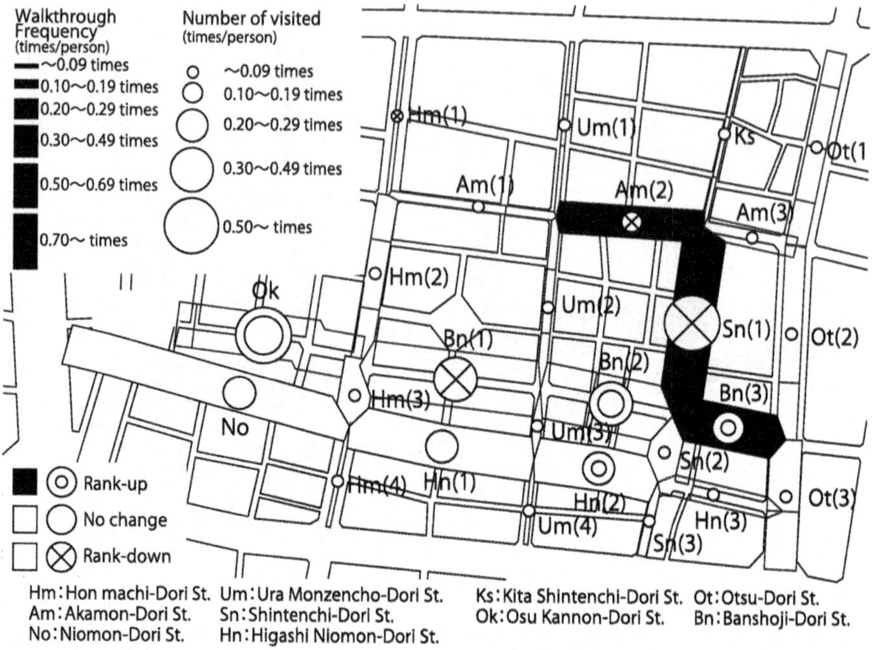

Fig. 1.5 Spatial distribution of 2008 walkthrough frequency and shop-visit numbers (Takeuchi et al. 2011)

0.43) and Shintenchi Dori St. (1) (0.88–>0.50). In 2003, the highest location value was 0.8; however, in 2008, it was as low as 0.6. This represents a decrease in the overall value and a decentralization of the shops visited. The decentralization of walkthrough frequency/shop-visit frequency from Bansho-ji (1) to Bansho-ji (2) and (3) may be attributed to the opening of a large shopping facility complex on the north side of location (1) in December 2003. The decrease in walkthrough frequency/shop-visit frequency in both Akamon Dori St. and Shintenchi Dori St. may be due to the many electronics shops that were located along Akamon Dori St., Shintenchi Dori St., Uramonzencho Dori St., and neighboring streets. In 2003, they were greatly reduced in number along Uramonzencho Dori St., while the number of tenants of electronics shops in the shopping facility complex on the west side of Akamon Dori St. decreased by half. The continuity of the localization of this shop type in the area was lost, which is assumed to have resulted in a decrease in walk-around shop-around behavior (Takeuchi et al. 2011).

1.3.4 The 2013 Survey

In the 2013 survey results, walkthrough frequency was highest in Banshoji Dori St. (1) (0.77), Banshoji Dori St. (2) (0.76), Osu Kannon Dori St., and Shintenchi

Dori St. (1) (0.63), respectively. In comparison with the previous survey, Shintenchi Dori St. (1) increased by 10%, which resulted in a change in order for Banshoji Dori St. (1) (2). Additionally, Banshoji Dori St., Akamon Dori St., Uramonzencho Dori St., Shintenchi Dori St. (1), and others increased, while Niomon Dori St., Higashi Niomon Dori St., Honmachi Dori St., and others decreased. District-wide, there was an increase to the north of Banshoji Dori St. and a decrease to the south.

Shop-visit frequency was highest in Shintenchi Dori St. (1) (0.60), Osu Kannon Dori St. (0.53), and Banshoji Dori St. (1) (0.49), respectively. From the previous survey, Shintenchi Dori St. (1) increased by 0.1 points, while Osu Kannon Dori St. decreased by 0.13 points. In addition, there was an increase in Akamon Dori St., Banshoji Dori St. (1), Uramonzencho Dori St., Kita-Shintenchi Dori St., Shintenchi Dori St. (1), and others, while Osu Kannon Dori St., Banshoji Dori St. (3), Higashi Niomon Dori St., Honmachi Dori St., and others decreased. There was an increase north of Akamon Dori St., Shintenchi Dori St. (1), and others, while there was a decrease south of the district centering on Higashi Niomon Dori St. (See also Chap. 2).

1.3.5 The 2018 Survey

Figure 1.6 contains the walkthrough frequency of each segment in the commercial districts and the shop-visit frequency per person from the 2018 survey. The walkthrough frequency was highest in Banshoji Dori St. (2) (0.76), Banshoji Dori St. (3) (0.69), and Banshoji Dori St. (1) (0.63), respectively. The most significant change from the last survey was the 12.5% increase in Banshoji Dori St. (3). The walkthrough frequency in the other segments generally decreased.

The order of shop-visit frequency per person was Shintenchi Dori St. (1) (0.47), Banshoji Dori St. (2) (0.40), and Osu Kannon Dori St. (0.37), respectively. In comparison with the 2013 survey, there was an increase in Otsu Dori Ave. (1), (2), and (3), but Banshoji Dori St. (1) (0.24), Osu Kannon Dori St. (0.16), and others all decreased. It is worth noting that Otsu Dori Ave., which was added to the survey area in 2008, was originally a pass-through segment street, but particularly on Otsu Dori Ave. (1), (2), and (3), both the walkthrough frequency and shop-visit frequency continued to increase across the 10-year period (Inagaki et al. 2020).

1.3.6 Shop-Around Corridor Patterns

Shop-around corridor patterns are unique to the Osu district. In commercial district clusters, corridor patterns that are passed through by many district visitors appear in a self-organized manner. There are many such corridors in the Osu district where the commercial districts are arranged in a grid-like pattern. The merits of each pattern arise from the minute differences in the locational balance of popular shops in each

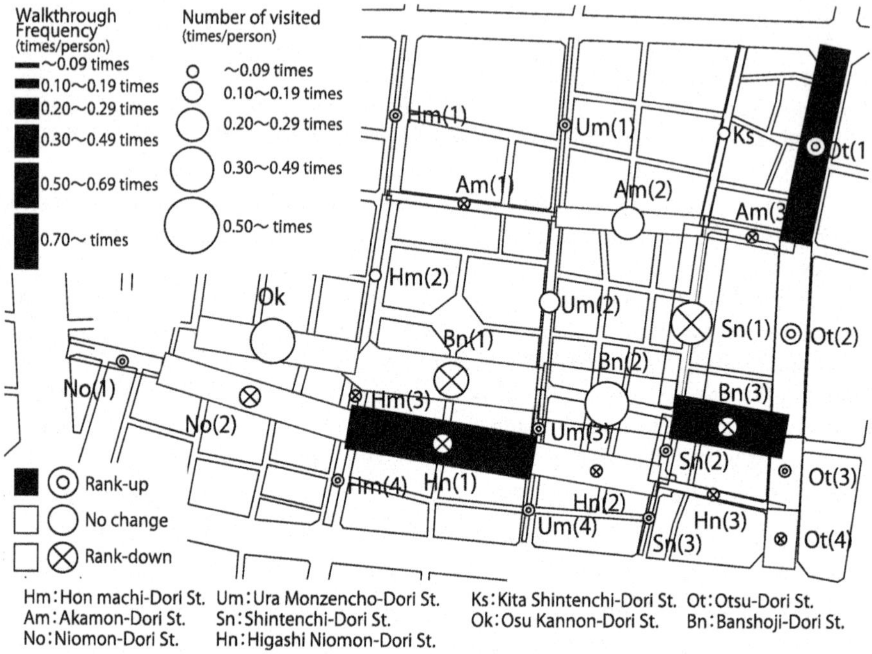

Fig. 1.6 Spatial distribution of 2018 walkthrough frequency and shop-visit numbers (Inagaki et al. 2020)

area. Additionally, while dramatic changes are observed in those merits, the balance of shop location is changed by the emergence among visitors of small groups with particular preferences, which results in the transformation of corridor patterns.

Walkthrough frequency data can be used to depict shop-around corridor patterns. The data is based on shop-around behavior because road links with a high walkthrough frequency can easily link the intersecting nodes. To summarize these trends, in the 1998 and 2003 surveys, a superior corridor pattern was evident that linked Akamon Dori St., Honmachi Dori St., Banshoji Dori St., and Shintenchi Dori St. (Loop-A); a dominant cluster of males with the purpose of visiting computer and other electronics shops was identified. Moreover, this corridor pattern was the route taken by users of the large car park that was established on Shintenchi Dori St. in 2003. However, in the 2008 survey, after the repair of the Higashi Niomon Dori St. arcade and at the time when there was a remarkable increase in food and drink shops on Osu Kannon Dori St., Niomon Dori St., and Higashi Niomon Dori St., precedence was taken by the corridor pattern of Banshoji Dori St., Osu Kannon Dori St., Niomon Dori St., and Higashi Niomon Dori St. (Loop-B). This was originally a round-trip road between Osu Kannon and Otsu Dori Ave. Below, the continuation of these trends based on the 2013 survey results is confirmed.

In the 2018 survey, the superior corridor pattern was Loop-B comprising Banshoji Dori St., Osu Kannon Dori St., Niomon Dori St., and Higashi Niomon

Dori St. with the addition of Shintenchi Dori St. (1) and (2). However, when making a detailed consideration of attributes and depending on the visitor group, the superior pattern was identified as Loop-A and a new manji pattern (Osu Kannon-Niomon Dori St.–Higashi Niomon Dori St.–Uramonzencho Dori St.–Akamon Dori St.–Otsu Dori Ave.) that appeared among young people.

1.4 Osu District Visitor-Cluster Analysis in the 2018 Survey

1.4.1 Summary of Visitor Configuration Analysis Based on a Cluster Analysis

Using a cluster analysis of the 2018 survey, the relationship between visitor-cluster characteristics and ratios in comparison with shop-around-behavior characteristics and survey data was studied.

A hierarchical cluster analysis was conducted using Ward's method and the square Euclidean distance. For the variables, the four indices of walking distance (X1), visit frequency (X2), shop-visit-plan ratio (X3), and shop-visit-purchase ratio (X4) were standardized and applied. Twenty-nine valid samples containing responses to the four indices were obtained. They were provided by respondents who completed Question 2 on the Osu district shop-around behavior questionnaire.

1.4.2 Attributes of Each Cluster

When performing the analysis, clusters were formed in the following six categories. Table 1.5 contains the average values of the attributes and variables as the characteristic indicators of each shop-around group.

1. **A Group ($n = 34$)**
   This group walks long distances, has a low visit frequency and has the lowest shop-visit-plan/-purchase ratio. This group also has an average number of companions and an average age. Accordingly, the group was termed low-frequency window-shopper.
2. **B Group ($n = 24$)**
   This group has the highest shop-visit-plan ratio and a relatively long walking distance. Its shop-visit-purchase ratio and visit frequency are average. The number of companions is also average, but it has a relatively young age range as it has the second-lowest average age. For these reasons, the group was named multipurpose shopper.
3. **C Group ($n = 44$)**
   This group has the lowest visit frequency, a low shop-visit-plan ratio, the longest walking distance, and a high shop-visit-purchase ratio. It also has the highest

Table 1.5 Characteristics of visitor clusters

	A group	B group	C group	D group	E group	F group	Total
Nickname	Low-frequency window-shopper	Multipurpose shopper	First-visit tourist	Local shopper	Shop-around lover	Single-purpose shopper	
(X 1) Average walking distance (m)	1906.4	2019.0	2322.7	862.7	1295.8	952.8	1559.9
(X 2) Move of visit frequency category	Once a year or more	Once a year or more	Once a year or more	Once a week or more	Once a week or more	Once a month or more	Once a year or more
(X 3) Ratio planned visit (%)	30.3	92.4	32.5	90.9	28.6	73.2	58.0
(X 4) Purchase ratio per store visit (%)	25.2	62.1	83.5	100.0	59.2	38.5	61.4
Composition (%)	14.8%	10.5%	19.2%	21.8%	14.0%	19.7%	100%
Number of samples (n)	34	24	44	50	32	45	229
Average age (age)	42.9	40.4	37.7	50.0	43.8	42.4	42.9
Average number of companion (person)	1.7	1.9	2.3	1.6	1.8	1.6	1.8

number of companions and the lowest average age, so it is thought to be a visitor cluster mainly comprising families with children. Accordingly, this group was termed first-visit tourists.

4. **D Group ($n = 50$)**
   This group has the highest shop-visit-plan/-purchase ratio, a high visit frequency, and the shortest walking distance. In addition, it has the highest average age and the lowest number of companions, so they are thought to be shoppers who live in or around Osu. For these reasons, the group was named local shoppers.
5. **E Group ($n = 32$)**
   This group has the highest visit frequency, and the lowest shop-visit-plan ratio. The walking distance and shop-visit-purchase ratio are average. It has average numbers of companions and an average age. For these reasons, this group was referred to as shop-around lover.
6. **F Group ($n = 45$)**
   This group has a high visit frequency and shop-visit-plan ratio and a lower ranking for both walking distance and shop-visit-purchase ratio. In addition, the group has an average age, and the number of companions is the lowest. For these reasons, this group was termed single-purpose shopper.

It is evident that groups can be configured such that there is an obvious disparity in the behavior characteristics of each unit. These characterize the shop-around behavior of visitor clusters based on external attributes. There is no extreme bias in the number of samples between groups, so a certain ratio within each group exists; this study has confirmed that there is a blend of visitor clusters with diverse objectives.

Each group's visitor cluster contains a certain proportion who are visiting for the first time with the purpose of sightseeing in the Osu commercial district. This supports the branding of this area as a tourist destination. Moreover, the visitor cluster comprises not only tourists but also local shoppers, those shopping in the commercial district, and those who enjoy shop-around behavior in itself; there are many demands for other than just sightseeing in the Osu commercial district. It is therefore confirmed that the district as a whole offers a number of shops for each visitor cluster.

1.5 Conclusion

This survey-centered study examined shop-around behavior in the Osu district. A questionnaire was used to gather data regarding visits to individual shops and the walking routes of individual visitors to the district. The survey combined data about shoppers' intentions in relation to the planning of their spatial actions in terms of shop-visit scheduling. Shop-around behavior is MPMS in nature, which is an essential data element for this analysis; it also featured in both Arakawa's walking-route redundancy analysis and Yoshida's ASSA modeling (see also Chap. 7).

Along with reviewing the results of surveys conducted between 2003 and 2018, the shop-around behavior of visitors to the district at each point in time was presented by means of walkthrough and visit frequencies. The spatial distribution of shop positioning through shop-type-configuration-survey data was also shown. Using individual shop data from the configuration survey and the spatial distribution of shop specialization and agglomeration in the district (which was obtained by aggregating each commercial district segment), it was possible to conduct an analysis of the micro-level interactions between shops and visitors by comparing individual visitor data. Now we can produce analyses of the Osu district's dynamics as panel data for 2003, 2008, 2013, and 2018.

Furthermore, the research explained the analysis of visitor clusters in the 2018 survey. Visitor clusters were categorized as low-frequency window-shopper, multipurpose shopper, first-visit tourist, local shopper, shop-around lover, and single-purpose shopper. In Chap. 2, Kobayashi et al. conducted a factor analysis based on the shop-efficiency ratio (number of visitors per shop) by measuring the transition between shops. It is expected that by subdividing visitor clusters, a more detailed analysis can be conducted to determine the effect of visitor clusters on shop-type configurations.

References

Arakawa M, Kaneda T (2002) Analyses on redundancy of shop-around behavior in CBD. J Archit Plan AIJ 556:227–233

Hayashi K, Ikemoto M, Kaneda T, Koyama Y, Nakamura J et al (2013) Actual surveys on Visitor's shop-around behaviors and vertical floor use in Akihabara, Tokyo. AIJ J Technol Des 19 (41):315–319

Inagaki S, Kaneda T, Takeuchi K, Ito R (2020) A study on characteristics of shopping street complex district visitors from viewpoint of shop-around behaviors—a survey of Osu district, Nagoya in 2018. In: Proceedings of Tokai Chapter Architectural Research Meeting

Ito A (1998) A study on shop-around behaviors in Osu District—comparison of action properties by the difference of gender, the age group. Graduation Thesis, Nagoya Institute of Technology

Kawaguchi S, Iki K (2016) Comparative analysis on rambling activities of visitors into the City Center. J Archit Plan AIJ 81(719):101–108

Kitamura A, Yamada T, Misaka T, Kaneda T (2003) The study of middle or advanced age female layer rerated to excursion behavior in the Center of the City: based on a questionnaire survey for the theaters of the Misono theater. In: AIJ, Summaries of Technical papers of Annual meeting, 2003, pp 1027–1028

Misaka T, Kaneda T (2003) The research related to the characteristics of excursion behavior in the Center of the City of the theater visitor—a case study on shin-Nagoya musical theater, Fushimi-Osamekakyo District in Nagoya. In: AIJ, Summaries of Technical papers of Annual meeting, 2003, pp 909–910

Misaka T, .Kaneda T (2005) An analysis on characteristics of theater visitors' shop around behaviors inside CBD (Central Business District)—a case study on Shin-Nagoya Musical Theater, Nagoya CBD. AIJ J Technol Des, 11(21): 469–474

Misaka T, Kaneda T (2006) An analysis on characteristics of art Museum visitors' shop around behaviors—a case study on the Nagoya City Museum. Stud Reg Sci 36(1):211–221

Misaka T, Kaneda T, Yamada S (2001) A study on revitalizing effects on downtown redevelopment project from the standpoint of walk-around behavior analysis: a case study on Fushimi-Nayabashi District, Nagoya. In: AIJ, Summaries of Technical papers of Annual meeting, 2001, pp 279–280

Misaka T, Oiwa Y, .Zheng D, Yoshida T, Kaneda T (2010) An analysis on characteristics of Visitors' shop-around Behaviours in commercial facility complex—a case study on Asunal Kanayama, Nagoya. Stud Reg Sci, 40(2): 511–525

Oiwa Y, Yamada T, Misaka A, .Kaneda T (2005) A translation analysis of Shopping District from the view point of Vistors' shop around behaviors -a case study of Osu District, Nagoya. AIJ J Technol Des, 11,(22):469–474

Saito S (1984) Construction of a non-aggregated multi-level Hough model considering the excursion between consumers. Plann Admin 13:73–82

Saito S (1998) A disaggregate Markov shop-around model to forecast sales of retail establishments based on the frequency of Shoppers' visits: its application to City Center retail environment at Kitakyusyu City. Pap City Plann 33:349–354

Takeuchi M, Yoshida T, .Kaneda T (2011) A translation analysis of shopping street complex district from the view point of shop around behaviors—a survey of Osu District, Nagoya in 2008, J Archit Plan AIJ, 76(660):361–368

Timmermans H (ed) (2005) Progress in activity-based analysis. Elsevier, Amsterdam

Timmermans H (ed) (2009) Pedestrian behaviors; models, data collection and applications. Emerald, London

Yamamoto T (1998) An analysis of attractive structure of Osu District by the Questionary survey. Graduation Thesis, Nagoya Institute of Technology

Chapter 2
Analyses on Transition Factors of Shop Tenants Inside Osu Shopping District

Yoshiyuki Kobayashi, Masaya Harasaki, and Toshiyuki Kaneda

Abstract The Osu shopping district in Nagoya has adapted to changes in the external environment by altering the configuration of its shops by the type of products they sell. This chapter looks at a 2013 survey centered on shop-around behaviors and the configuration of shops in the district and compares it with an earlier one conducted in 2008. By analyzing the changes across the two surveys, a dynamic model of factors behind those changes emerges, illustrated in this chapter through a series of tables and charts. Shop distribution in the Osu district was recorded via a field study, and shop tenants were invited to complete a questionnaire on such factors as visitor attributes, their walking routes, and their purposes of visit. The findings included: (1) patterns of change in shop types during this period; (2) a spatial shift in the visitors' behaviors; and (3) crowding indicators as potential factors behind shop tenant changes. Additionally, a relationship between the amount of walking around by visitors and the density of shops was confirmed.

2.1 Research Background and Objectives

Since the 1980s, Japanese city centers have been declining due to the rise of motorization and the expansion of large-scale commercial shops in city suburbs. Policies and plans have been drawn up to revitalize city centers, including the enactment and subsequent revision of the Three Town Development Laws.

Y. Kobayashi
Nagoya City Office, Sannomaru, Naka, Nagoya, Japan
e-mail: y.kobayashi.13@city.nagoya.lg.jp

M. Harasaki
Asahi Kasei Homes Corporation, Chiyoda-ku, Japan

T. Kaneda (✉)
Graduate School of Engineering, Nagoya Institute of Technology, Gokiso, Showa, Nagoya, Japan
e-mail: kaneda@nitech.ac.jp

T. Kaneda (ed.), *Downtown Dynamics*, Agent-Based Social Systems 16,
https://doi.org/10.1007/978-4-431-54901-7_2

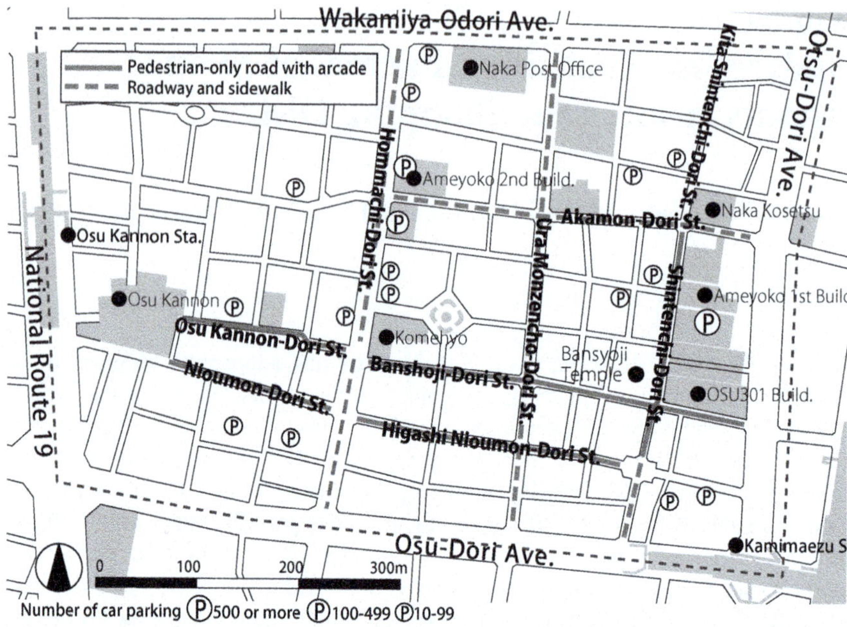

Fig. 2.1 Composition of the nine main streets in Osu district, Nagoya (2013)

The Osu district (Fig. 2.1), located to the south of Nagoya City center, is an area of approximately 30 ha enclosed by Wakamiya-Odori Ave., Otsu-Dori Ave., Osu--Dori Ave., and National Route 19 (Fushimi-Dori Ave.), with the Osu Kannon underground station to the west and the Kamimaezu underground station on the south-eastern side. This is a substantial shopping district with multiple streets containing mainly small independent retailers. It is often crowded with visitors coming from all points of the compass, attracted by its unique character, and vibrant mix of people.

When considering policies for the revitalization of city centers, it is important that research is conducted into the source of the adaptability that drives customer–retailer interaction in connection with changing contemporary needs and environments. With this as the focus, we observed configurations of shopping zones according to shop type and visitor shop-around behaviors in Osu in the years 1998, 2003, 2008, and 2013 (Oiwa et al. 2005; Takeuchi et al. 2011).

By comparing our 2013 data with the survey in 2008 (Takeuchi et al. 2011), the characteristics of temporal changes in shop type configurations and visitor shop-around behaviors were clarified. Based on our analysis, dynamic factors behind shop tenant transitions in Osu emerged.

Previous studies include research into the configuration of shops by retailing category/type in commercial hubs in city centers (Miyamoto and Yuzawa 2004; Yamada et al. 2009), visitor shop-around behaviors (Arakawa and Kaneda 2002; Takahashi et al. 2005; Park and Satoh 2006), and research into the mutual interactions between shoppers and shopkeepers (Oiwa et al. 2005; Takeuchi et al. 2011; Hayashi et al. 2013), but no reports have yet analyzed shop tenant transitional factors with a focus on visitor shop-around behavior.

2.2 Shop Configuration by Category/Type in the Osu District

2.2.1 A Comparison Between 2008 and 2013

All of the shops within the nine shopping street promotion associations (hereafter called nine main streets) were within the target area of our research into the configuration of shop by retailing category/type in the Osu district and on the west side of Otsu-Dori Ave. The research was conducted through field surveys, and by reference to residential maps from Zenrin Co. Ltd., targeting first floor, second floor and above, and basement levels one and below. For this study, shop types were categorized into a total of 10, namely *food and drink*, *groceries*, *variety*, *furniture*, *electrical goods*, *clothing*, *temples/parks*, *public car parks*, *vacant shops/residences/offices/private car parks/vacant lots*, and *other*. In the more complex commercial facilities, to simplify the data, the shop type was designated as that of the establishment with the largest floor area. Further, shops occupying the first floor of a complex commercial facility with a separate entrance leading directly onto the street and whose business hours were not defined by the management of the commercial facility were counted as a single shop, separate from that facility.

For this study, the 10 types of establishment were further grouped into four main types: (1) *food and drink/groceries*, (2) *variety/furniture*, (3) *electrical goods*, and (4) *clothing*, with the addition of (5) *temple/park* and (6) *other*, making a total of six categories. The large recorded numbers of visitors to *public car parks* and *vacant shops/residences/offices/private car parks/vacant lots* were grouped under *other* because it was not always possible to establish the purpose of their visits.

Figure 2.2 shows the configuration of shops by type along the nine main streets except for the west side of Otsu-Dori Ave. The total number of shops grew during the 5-year period with the total of the four main types increasing from 457 to 470 shops. There was no change in the order of category size within the shop composition ratio, with the largest group being *clothing* and the smallest being *variety/furniture*. *Food and drink/groceries* was the category with the largest increase in shops, which may have been influenced by the trend toward walking and eating on the street that is characteristic of this district, while *clothing* had the greatest decrease, which may have been influenced by the boom in second-hand clothes. On the west side of Otsu-Dori Ave., where the survey began in 2008, the number of shops appeared to have doubled by 2013 (from 24 to 42). However, this number may be partly due to differences in research methods between the two surveys, so the west part of Otsu-Dori Ave. was excluded from the study's analysis.

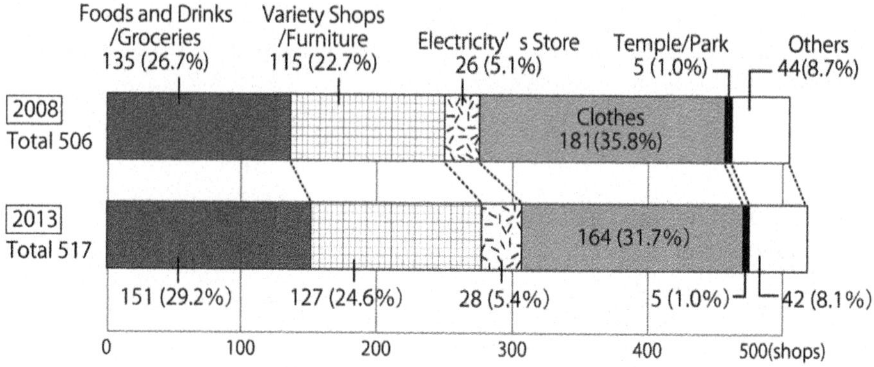

Fig. 2.2 Shop type configurations along the nine main streets

2.2.2 *Spatial Distribution of Shops and Configuration by Retail Category on Each Street*

Figure 2.3 shows the spatial distribution of shops along the nine main streets and west Otsu-Dori Ave., and Fig. 2.4 shows the configuration of shop types on each of the nine main streets. The total number of establishments increased over the 5 years on six of the streets, including Akamon-Dori St. and Osu Kannon-Dori St., while

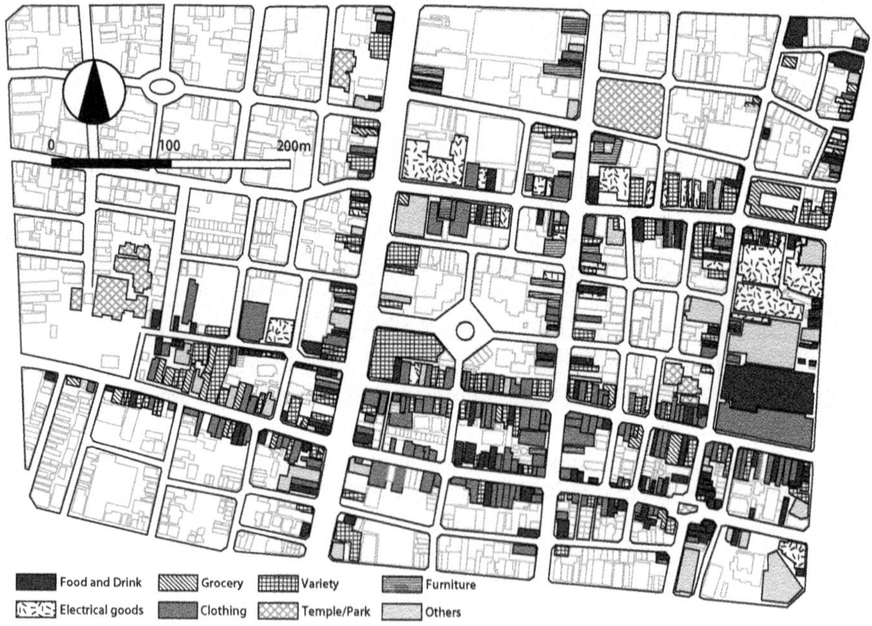

Fig. 2.3 Spatial distribution of shop locations along the nine main streets

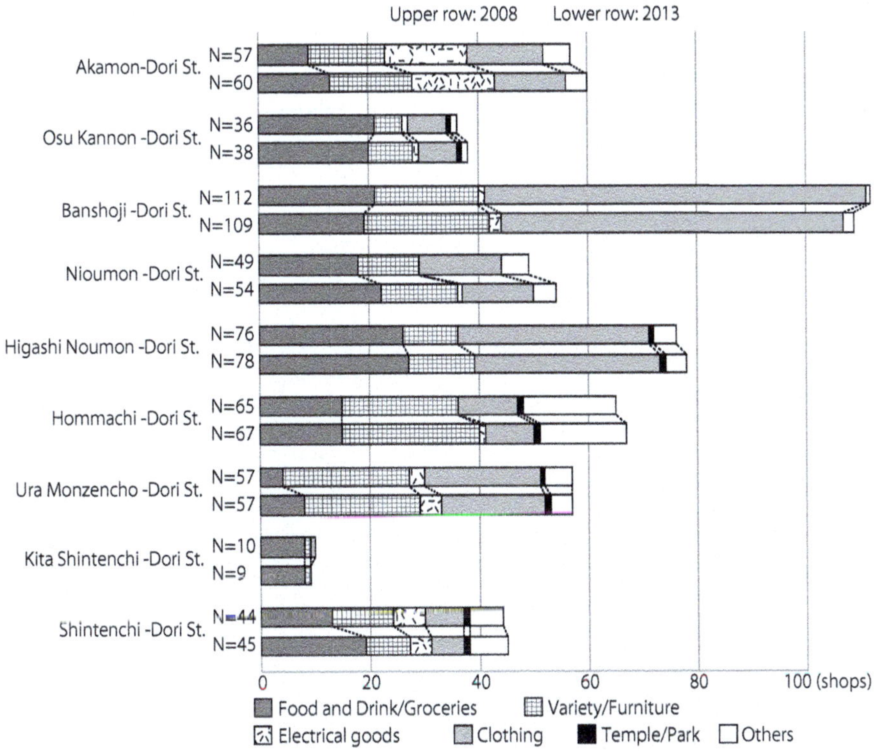

Fig. 2.4 Shop type configurations for each of the nine main streets

they decreased on Banshoji-Dori St. and Kita Shintenchi-Dori St.; there was no change in the total number of shops on Ura Monzencho-Dori St. Looking at the shop type configuration on each street, it was found that the types of shops tended to gather together by street; for example, *electrical goods* on Akamon-Dori St., *food and drink/groceries* on Osu Kannon-Dori St., *clothing* on Banshoji Dori St., and *variety/furniture* on Ura Monzencho-Dori St. While retaining those characteristics, *food and drink/groceries* also increased on Akamon-Dori St. (from 9 to 13 shops), Ura Monzencho-Dori St. (from 4 to 8 shops), and Shintenchi-Dori St. (from 13 to 19 shops), while *variety/furniture* increased on Osu Kannon-Dori St. (from 5 to 8 shops) and Banshoji Dori St. (19 to 23 shops). However, there was a general decrease in shops in the *clothing* category, particularly on Banshoji-Dori St. (from 70 to 63 shops).

Table 2.1 Shop tenant transitions from 2008 to 2013 (including mergers/separations along the nine main streets)

		Number of shops								
		Food and drink/groceries	Variety/furniture	Electrical goods	Clothing	Others	Vacant shops, etc.	Loss	Total (C)	2008 number of shops (D)
Number of shops	Food and drink/groceries	(95) 29	4	1	4	0	5	2	(95) 45	135
	Variety/furniture	9	(84) 14	4	4	0	4	0	(84) 35	115
	Electrical goods	0	3	(17) 4	1	1	0	0	(17) 9	26
	Clothing	8	18	1	(121) 28	3	7	0	(121) 65	181
	Others	4	1	0	2	(33) 4	1	1	(33) 13	44
	Vacant shops, etc.	7	6	3	4	1	(80) 7	0	(80) 28	105
	New	0	0	0	0	0	0		0	
Total (A)		(95) 57	(84) 46	(17) 13	(121) 43	(33) 9	(81) 24	3	(430) 195	607
2013 number of shops (B)		151	127	28	164	42	103		615	

( ): Not changed

2.2.3 Changes in Shop Tenants

Table 2.1 shows the changes in shop tenants along the nine main streets. Total (A) shows the number of new shop openings including mergers, while Total (C) shows the number of shops closing including separations. The influence of mergers and separations is minor; therefore, for the number of shop openings in this research, the opening ratio (number of new shop openings/(B)) was defined by subtracting the number of shops that had not changed since 2013 (B). Regarding the number of shops closing down, the ratio (number of shops closing/(D)) was also defined by subtracting the number of shops that had not changed since 2013 (D). The scope did not include *temples/parks*, which did not change.

The category of *food and drink/groceries* had the most new openings (56 shops), while *clothing* had the most shops closing down (60 shops). Shops that closed down were often replaced by another in the same retail category, but there were notable shop type changes from *variety/furniture* to *food and drink/groceries* (9) and from *clothing* to *variety/furniture* (18). It was also found that there were conversions from *clothing* to *variety/furniture* and from *variety/furniture* and *clothing* to *food and drink/groceries* within the Osu district.

Looking at the changes on each street shown in Fig. 2.4, there was notable replacement from *variety/furniture* to *food and drink/groceries* on Akamon-Dori St. (two out of four replacement shops) and Ura Monzencho-Dori St. (three out of eight), and from *clothing* to *food and drink/groceries* on Shintenchi-Dori St. (three out of three); while on Shintenchi-Dori St., there was increased specialization in *food and drink/groceries.* Additionally, on these three streets, as well as Banshoji-Dori St. and Hon machi-Dori St. Street, *variety/furniture* shops were replaced by *electrical goods*, while on Akamon-Dori St., *electrical goods* were replaced by *variety/furniture* shops (2 out of 5 replacement shops). On Osu Kannon-Dori St. (two out of four) and Banshoji-Dori St. (6 out of 15), *clothing* shops were replaced by *variety/furniture* shops.

2.3 Shop-Around Behavior in the Osu District

2.3.1 Summary of the Shop-Around Behavior Survey

At five distribution points in the target district (defined on the basis of a simple passage volume survey conducted beforehand), a questionnaire-based survey was conducted. The number of questionnaires distributed was 1346, of which 316 valid responses were collected (a collection ratio of 23.5%). The questionnaire requested details on visitor attributes, the walking route in the Osu district on that day, the shops visited, whether the visits were planned, and more. Answers were obtained on gender, age, type and number of companions, means of transport, and more. Respondents used a map to show the walking route and the location as well as the order of shops visited by them. Questions were also asked on the shop names, duration of the visit, whether the visit was pre-planned, and whether products or services were purchased.

2.3.2 *Attributes of Respondents*

Table 2.2 shows a summary of visitor attributes at two points in time: 2008 and 2013. In the shop-around behavior survey, although the questionnaire covered nine visitor attributes, namely, gender, age, occupation, address, means of transport, number of companions, type of companions, location immediately before/after, and frequency of visits, the basic attributes used in the analysis conducted in this survey were gender and age.

Looking at the visitor composition ratio in terms of gender, males was slightly higher than females at 53%. Compared with past surveys, females had been on the increase until 2008, but males had increased by 2013, showing a marked change in the gender ratio. In terms of age, the highest category of shoppers was *50 years old or above*, followed by *30 to 49 years old*, and *29 years old or below*, which shows a decrease in visitors between the ages of 30 and 49 years and an increase in the *50 years old or above* category since 2008.

2.3.3 *Shop-Around Behavior*

In 2013, the average walking distance of respondents was 1.49 km, the average visit duration was 182 min, and the average shop visit count was 4.21. In comparison with the previous survey, the walking distance had increased by 0.03 km and the visit duration by 15 min, while the number of shops visited decreased by 0.26. In comparison with 10 years ago, the visit duration and walking distance were increasing and the shop visit count was decreasing; however, from a commercial point of view, it is preferable for the number of shops visited per person to increase.

Focusing on differences in shop-around behavior by gender, the ratios for visit duration, walking distance, and shop visit count were higher among females than males, which shows no change from previous surveys. Looking at age groups, shop-around behavior tended to become livelier according to the higher age of the group, which is a marked change from previous surveys.

Table 2.2 Main characteristics of the shop-around behavior survey by attributes

		Number of visitors	Ratio [%]	An average visit duration [min]	An average walking distance [km]	An average shop visit count
2013	Total	295		182	1.49	4.21
Gender	Male	155	53	174	1.40*	3.98
	Female	140	47	190	1.57	4.45
Age	29 years old or below	60	20	163	1.39	3.93
	30–40 years old	111	38	184	1.52	4.49
	50 years old or above	124	42	189	1.50	4.09
Address	Naka Ward, Nagoya City	35	12	177	1.55	3.31
	Nagoya City	118	40	164	1.42	4.04
	Aichi Pref.	96	33	199	1.45	4.54
	Others	45	15	183	1.70	4.58
Visit frequency	Every day	13	4	284	1.55	3.00
	A few times a week	42	14	183	1.22	3.90
	A few times a month	105	36	169*	1.49	4.57
	A few times a year	97	33	178	1.51	4.17
	Hardly come	37	13	191	1.63	3.88
Number of companions	1	107	36	165	1.32	3.26
	2	127	43	192	1.58**	4.52**
	3 or more	35	12	185	1.63	2.53
Types of companions	Friends	47	16	215	1.50	4.60
	Family	54	18	181	1.57	4.71
	Couple	85	29	186	1.62*	4.54*
	Others	13	4	177	1.73	3.80
	Alone	107	36	165	1.32	3.26
Means of transport	Subway or bus	153	52	193	1.42	4.08
	Car	67	23	192*	1.63	5.10**
	Walking or bicycle	64	22	144	1.47	3.52
	Others	10	3	184	1.72	5.64
2008	Total	582	0	167	1.46	4.47

(continued)

Table 2.2 (continued)

		Number of visitors	Ratio [%]	An average visit duration [min]	An average walking distance [km]	An average shop visit count
Gender	Male	243	42	162	1.40	4.27
	Female	332	58	170	1.51	4.63
Age	29 years old or below	118	23	188	1.54	4.61
	30–40 years old	255	50	172**	1.48	4.81**
	50 years old or above	208	41	148	1.39	3.98
Address	Naka Ward, Nagoya City	64	11	138	1.28	3.36
	Nagoya City	271	47	159**	1.42**	4.55**
	Aichi Pref.	164	29	186	1.52	4.57
	Others	76	13	169	1.61	5.10
Visit frequency	Every day	26	46	156	1.42	2.28
	A few times a week	82	14	178	1.30	4.05
	A few times a month	174	31	159	1.45**	4.68**
	A few times a year	195	34	172	1.51	4.91
	Hardly come	91	16	166	1.60	4.12
Number of companions	1	159	28	136	1.28	3.88
	2	284	49	173**	1.57**	4.79**
	3 or more	132	23	186	1.45	4.50
Types of companions	Friends	100	18	205	1.59	5.12
	Family	166	30	174**	1.45**	4.68**
	Couple	133	24	152	1.60	4.59
	Alone	159	28	136	1.28	3.88
Means of transport	Subway or bus	3.16	55	182	1.46	4.71
	Car	154	27	156**	1.50	4.58**
	Walking or bicycle	95	16	134	1.36	3.54
	Others	11	19	176	1.56	5.10

($*P < 0.1$; $**P < 0.05$)

2.3.4 *Visitor Characteristics from the Viewpoint of the Visited Shop Types*

Table 2.3 shows the shop visit count (visits/person) by attribute/shop type. Overall, the most visited were, in order, *food and drink* (0.92), *variety/furniture* (0.82), *clothing* (0.63), and *electrical goods* (0.61).

Regarding shop type categories, the category of *used* from the 2008 survey was classified into each shop type, and the *complex facilities* category was re-organized into the largest type of shop tenant on each floor, so it could not be determined. However, in comparison with previous studies, the implication was that visits to *electrical goods* shops had increased in the group with attributes such as *male* and *alone*. Moreover, visits to *clothing* shops in the group with attributes such as *female* and *with friends* dropped from first place to third place, suggesting an overall decrease in visits to *clothing* shops.

2.3.5 *Ratios of Planned Visits to Shops*

Figure 2.5 shows the ratio of planned/unplanned shop visits by shop type. Overall, planned shop visits in the 2013 survey amounted to 45.5% of all shop visits, which is a decrease from the previous survey (50.6%). Further, the planned shop visit count per person was 1.91 visits per person, which was lower than the previous survey (2.21).

Looking at individual shop types, planned visits decreased in all shop types except in the *others* category, with *electrical goods* and *others*, each showing more planned visits than unplanned in both 2013 and 2008. When considering the increase in unplanned shop visits along with the fact that the visitor walking distance increased over the previous survey, there is a notable trend toward visiting shopping streets with a purpose but also to walking around within the district and visiting several shops.

2.3.6 *Ratio of Purchasing/Non-Purchasing*

Figure 2.6 shows the shop visit purchasing/non-purchasing ratio by shop type. Overall, 51.3% of shoppers made a purchase during their visit to a shop according to the 2013 survey, which was lower than in 2008 (56.4%). Additionally, the number of shops visited per purchase per person was 2.19, which was also a drop (from 2.34). Purchasing increased for *food and drink/groceries* and *variety/furniture*, while there was a decrease in purchasing of *electrical goods* and *clothing*. The overall decrease in purchasing suggests an increase in visitors with lower purchasing power,

Table 2.3 Numbers of shop visits (visits/person) by attribute/shop type

		N	1st		2nd		3rd	
2013	Total	286	Foods and drinks	0.92	Variety Shops/Furniture	0.82	Clothes	0.63
Gender	Male	149	Electricity's store	0.95	Foods and drinks	0.65	Variety Shops/furniture	0.60
	Female	137	Foods and drinks	1.12	Variety shops/furniture	0.99	Clothes	0.88
Age	29 years old or below	61	Temple/park	1.00	Clothes	0.85	Foods and drinks	0.82
	30–40 years old	106	Foods and drinks	1.12	Variety shops/furniture	0.94	Clothes	0.73
	50 years old or above	119	Foods and drinks	0.78	Electricity's store	0.62	Temple/park	0.62
Means of transport	Subway or bus	152	Foods and drinks	0.82	Variety shops/furniture	0.74	Electricity's store	0.67
	Car	61	Foods and drinks	1.41	Variety shops/furniture	1.0	Clothes	0.90
	Walking or bicycle	61	Foods and drinks	0.64	Clothes	0.57	Electricity's store	0.44
	Others	11	Variety shops/furniture	1.18		1.09	Foods and drinks	1.09
Number of companions	1	108	Electricity's store	0.90	Variety shops/furniture	0.67	Foods and drinks	0.57
	2	126	Foods and drinks	1.04	Clothes	0.83	Variety shops/furniture	0.83
	3 or more	51	Foods and drinks	1.31	Variety shops/furniture	1.00	Clothes	0.73
Types of companions	Friends	43	Electricity's store	0.93	Variety shops/furniture	0.81	Clothes	0.63
	Family	52	Foods and drinks	1.25	Variety shops/furniture	1.19	Variety Shops/furniture	0.98
	Couple	81	Foods and drinks	1.07	Clothes	0.90	Variety Shops/furniture	0.78
	Others	5	Foods and drinks	3.60	Temple/park	2.40	Others	1.60
	Alone	108	Electricity's store	0.83	Variety shops/furniture	0.65	Foods and drinks	0.58
2008	Total	565	Foods and drinks	0.94	Clothes	0.85	Used/complex facilities	0.65
Gender	Male	228	Used/complex facilities	0.84	Foods and drinks	0.79	Variety shops/furniture	0.63
	Female	328	Clothes	1.13	Foods and drinks	1.05	Variety shops/furniture	0.64
Age	29 years old or below	112	Clothes	1.29	Foods and drinks	0.93	Variety shops/furniture	0.76
	30–40 years old	252	Foods and drinks	1.10	Clothes	0.92	Variety shops/furniture	0.77
	50 years old or above	200	Foods and drinks	0.76	Used/complex facilities	0.66	Temple/park	0.64
Means of transport	Subway or bus	309	Foods and drinks	1.01	Clothes	1.00	Variety Shops/furniture	0.64
	Car	145	Foods and drinks	1.03	Clothes	0.93	Used/complex facilities	0.77
	Walking or bicycle	95	Variety shops/furniture	0.67	Foods and drinks	0.66	Groceries	0.56

Number of companions	1	155	Used/complex facilities	0.86	Variety shops/furniture	0.54	Clothes	0.52
	2	276	Foods and drinks	1.07	Clothes	0.97	Variety shops/furniture	0.73
	3 or more	125	Foods and drinks	1.23	Clothes	0.99	Variety shops/furniture	0.58
Types of companions	Friends	99	Clothes	1.27	Foods and drinks	1.21	Variety shops/furniture	0.86
	Family	161	Foods and drinks	1.20	Clothes	0.91	Variety Shops/furniture	0.62
	Couple	128	Foods and drinks	0.96	Clothes	0.88	Used/complex facilities	0.64
	Others	156	Used/complex facilities	0.86	Variety shops/furniture	0.54	Clothes	0.52

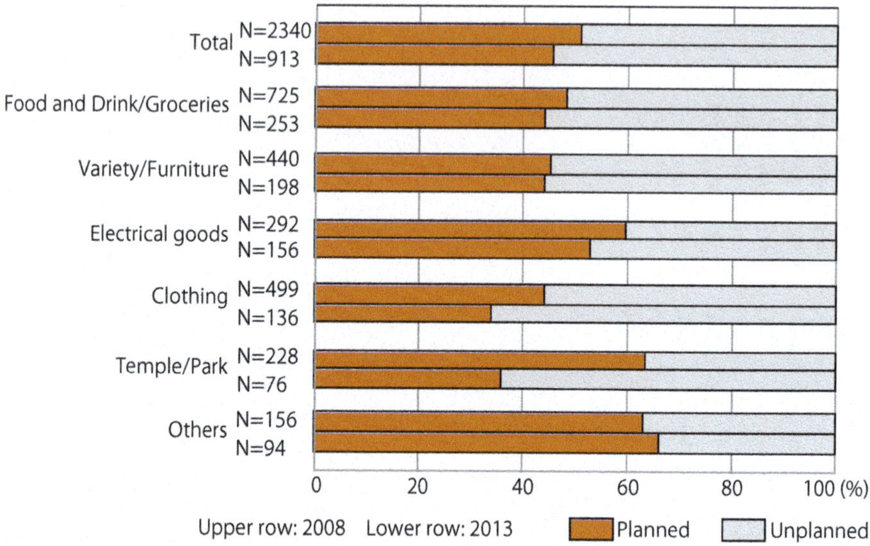

Fig. 2.5 Ratio of planned/unplanned visits by shop type

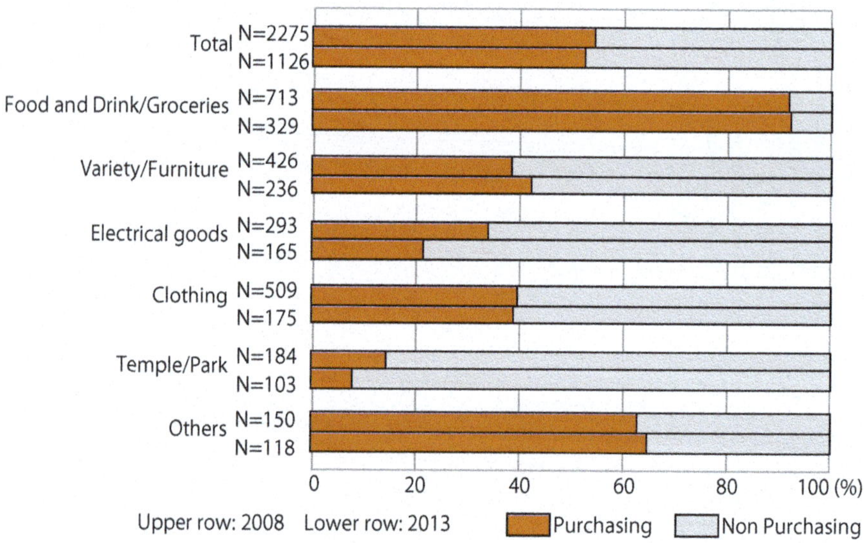

Fig. 2.6 Ratio of purchasing/non-purchasing by shop type

and it is thought that visitors may have become more cautious about purchasing *electrical goods*.

Figure 2.7 shows the ratio of planned and unplanned visits that involved a purchase. There was an increase in unplanned visits as well as for purchases during

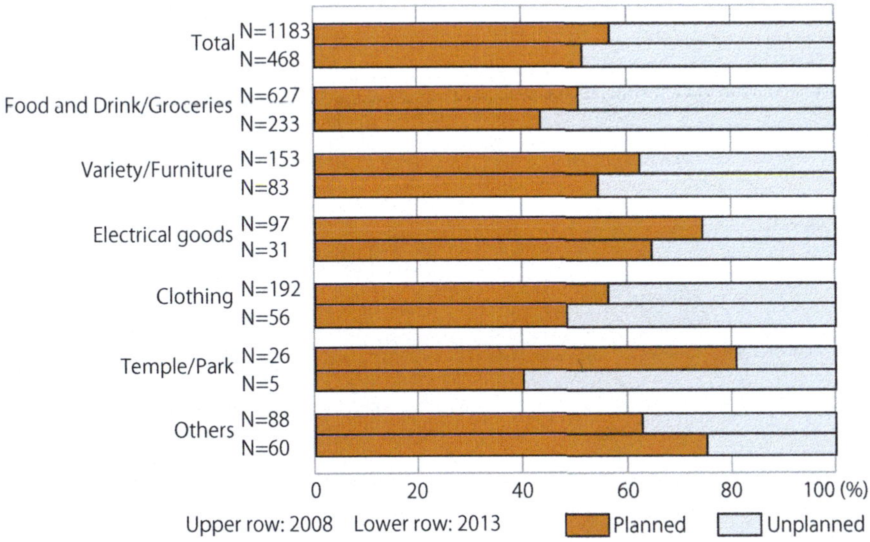

Fig. 2.7 Ratio of planned/unplanned visits that involved a purchase by shop type

those visits, which could suggest an increase in visitors who walk around looking at various shops, i.e., visitors who "shop around," before purchasing.

2.3.7 Spatial Distribution of Walk-Through Frequency and Shop Visit Count by Street

Figure 2.8 shows the walk-through frequency (total route numbers/sample number × 100) and the shop visit count per person (shop visit total/sample number) for each section of the shopping streets. The walk-through frequency was highest on Banshoji-Dori St. (1) at 77%, followed by Banshoji-Dori St. (2) at 76%, Osu Kannon-Dori St., and Shintenchi-Dori St. (1) at 63%. In comparison with the previous survey, Banshoji-Dori St. (1) and (2) switched ranks, while Shintenchi-Dori St. (1) increased by 10%. Looking at the district as a whole, there were increases in walk-through frequency on Banshoji-Dori St., Akamon-Dori St., Ura Monzencho-Dori St., and Shintenchi-Dori St. (1), while there were decreases on Nioumon-Dori St., Higashi-Nioumon-Dori St., and Hon machi-Dori St.

According to the 2013 survey, the number of shop visits per person was highest on Shintenchi-Dori St. (1) at 0.60 visits per person, followed by Osu Kannon-Dori St. at 0.53, and Banshoji-Dori St. (1) at 0.49. Compared with 2008, Shintenchi-Dori St. (1) had increased by 0.1 visits per person, while Osu Kannon-Dori St. had decreased by 0.13. Looking at the district as a whole, Akamon-Dori St., Banshoji-Dori St. (1), Ura Monzencho-Dori St., Kita Shintenchi-Dori St., and Shintenchi-Dori St. (1) all experienced an increase over the 5-year period, while there was a decrease

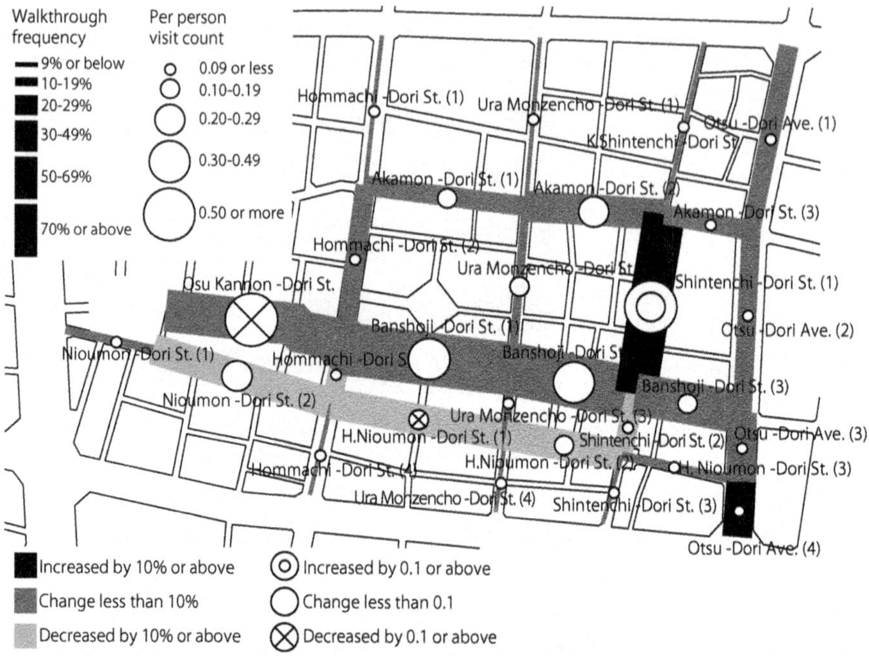

Fig. 2.8 Spatial distribution of walk-through frequency and shop visit count of the streets in 2013

on Osu Kannon-Dori St., Banshoji-Dori St. (3), Higashi-Niomon-Dori St., and Hon machi-Dori St. among others.

The increase in walk-throughs and visits on Akamon-Dori St., Shintenchi-Dori St., and Ura Monzencho-Dori St. may be due to the substitution of *food and drinks/ grocery* shops for *electrical goods* in those locations. Routes and visits on Higashi-Niomon-Dori St. and Hon machi-Dori St. tended to decrease, which may also be due to the decline in *food and drink/grocery* and *clothing* shops. One of the reasons for the decrease in visits to Osu Kannon-Dori St. may be due to the closure of a large-scale shop.

2.4 Factors Behind Shop Tenant Transitions in the Osu District

2.4.1 Analytical Framework of Factors Behind Shop Tenant Transitions

Figure 2.9 shows a model for analyzing the dynamic factors behind shop tenant transitions. The visit count per shop (per person visit count/number of shops) in the model shows the customer concentration based on the number of visits per person

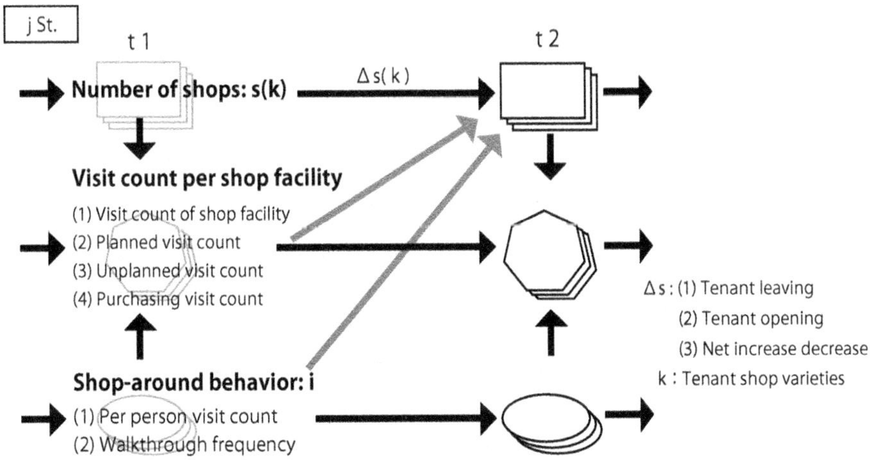

Fig. 2.9 Dynamic model of shop tenant transition factors

per shop, which is an indicator that affects shop management. In this chapter, using the visit counts per shop and the data on walk-through frequency from both surveys, we looked to identify a dynamic link with the patterns of shop tenant transitions. In this study, we considered the visit count per shop by shop type/street, as well as the number of planned/unplanned/purchasing/non-purchasing visits. Moreover, configuration changes in shop type were broken down into shop opening, shop closing, and net increase/decrease. Thus, a series of factors behind shop tenant transitions were derived, based on an analysis of each street. The analysis targeted only the four main shop types and did not include *temples/parks*, which did not change, or *others*, which comprised miscellaneous shop tenant varieties.

2.4.2 Walk-Through Frequency and Visit Count Per Shop by Street

The connection between the walk-through frequency based on visitor attributes (gender/age group) and the visit count per shop among the number of unplanned visits was analyzed for both 2008 and 2013. As a result, among all attributes, a correlation was found between the walk-through frequency and the number of visits per shop in the *food and drink/groceries*, *variety/furniture*, and *clothing* categories. The foremost results are shown in Fig. 2.10. This signifies a correlation between the number of pedestrians on the street and attracting responsive visitors.

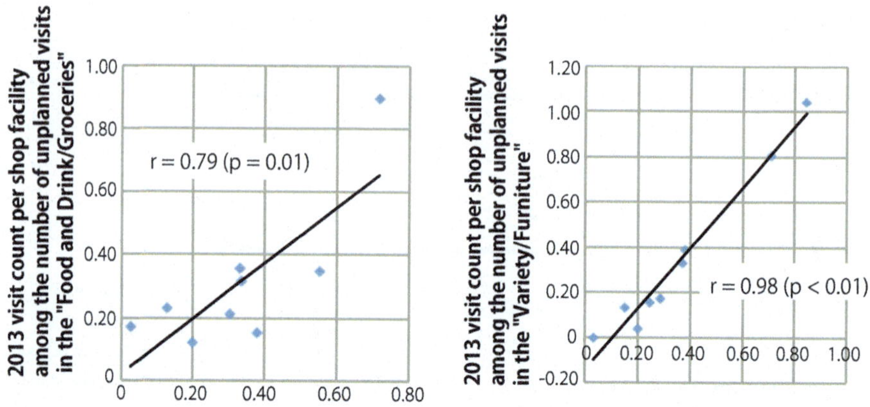

Fig. 2.10 Walk-through frequency and unplanned visit count per shop

2.4.3 Analysis of Dynamic Factors and Characteristics of Shop Type

Figures 2.11, 2.12, 2.13, 2.14, and 2.15 show the results of the analysis of the factors behind shop tenant transitions, changes in shop type configurations across the district, and visitor shop-around behavior for each street across a 5-year period, broken down by shop type. In the analysis using the 2008 visit count per shop, it was not possible to clarify the dynamic factors, so an explanation is provided here that focuses on the results of analyzing the visitor shop-around behavior index. In addition, for the specialization factors for the visit counts per shop, the counts for each street are divided by the average values for the respective shop type, so that a

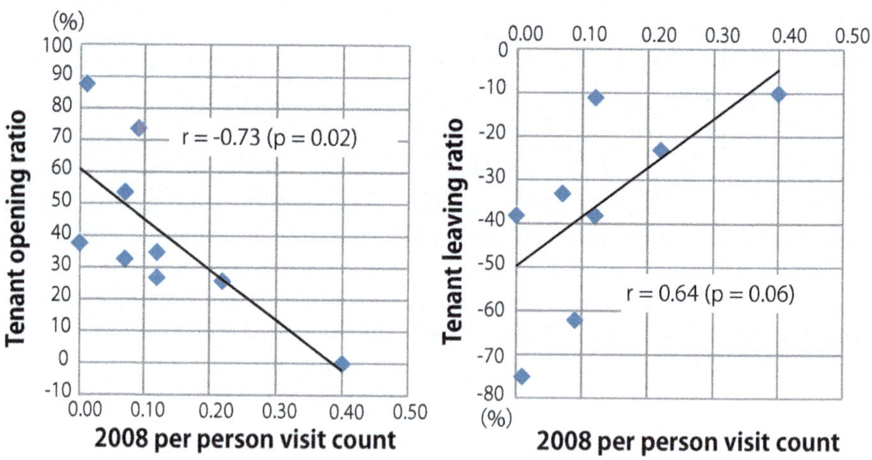

Fig. 2.11 Visit count and openings/closures ratio of food and drink/groceries

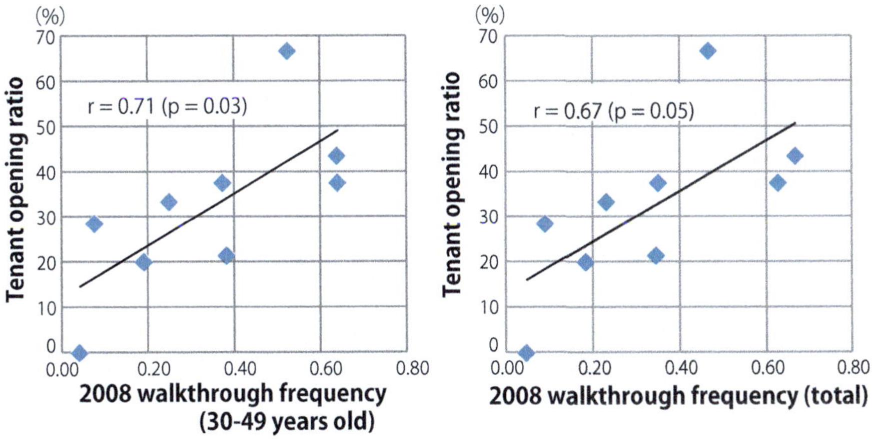

Fig. 2.12 Change across 5-year period in food and drink/groceries

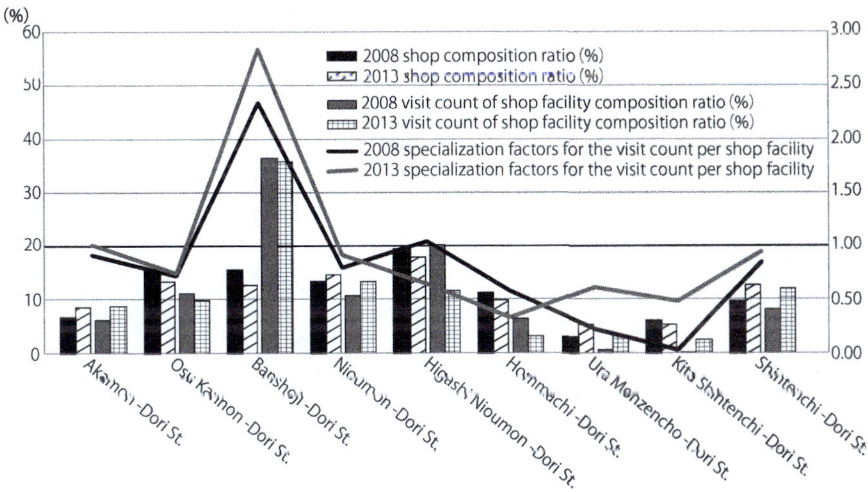

Fig. 2.13 Visit count and openings/closures ratio of variety shops/furniture

score of 1 or above denotes a greater-than-average ability to attract customers for each shop (Figs. 2.16 and 2.17).

In the *food and drink/groceries* category, there was a correlation between the visit count per person and the shop opening/closing down ratio, which may be interpreted as a trend for replacements to increase in intensity as the shop opening/closing down ratio increases according to fewer visits. There were new shop tenant openings in 7 out of 8 shops on Ura Monzencho-Dori St., 14 out of 19 shops on Shintenchi-Dori St., and 7 out of 13 shops on Akamon-Dori St., and the visit count per person increased alongside the composition ratio while the visit count per shop also tended to increase.

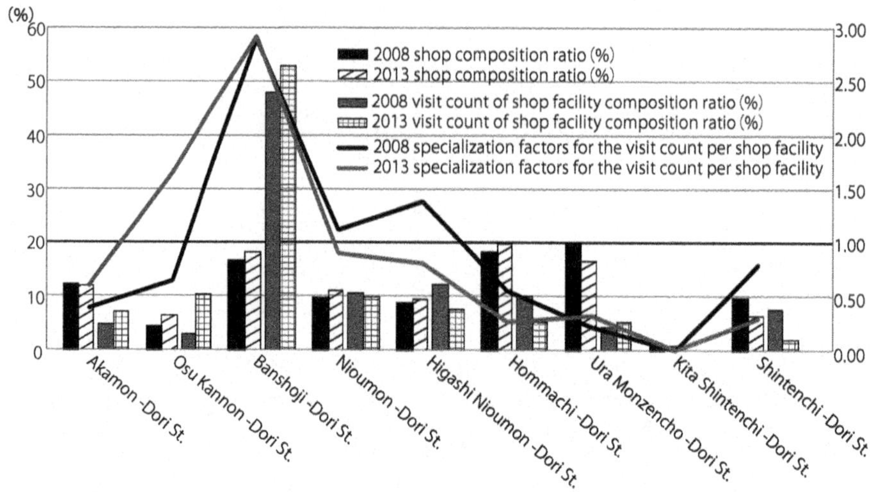

Fig. 2.14 Change across a 5-year period in variety shops/furniture

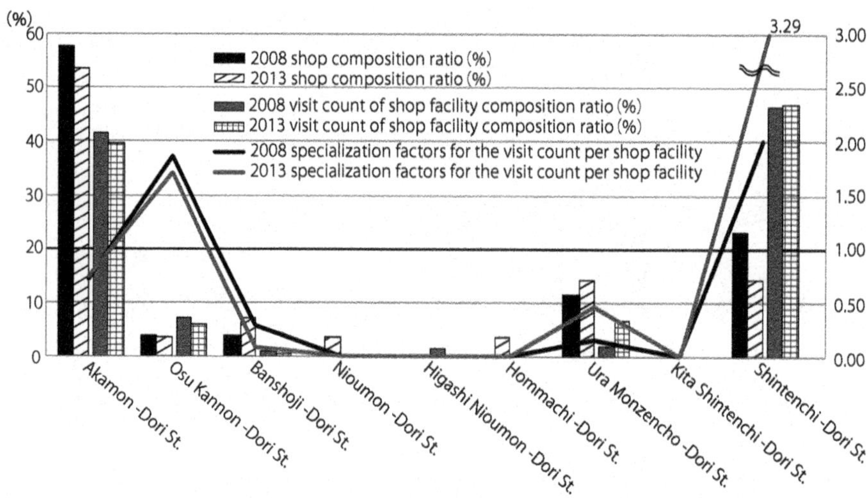

Fig. 2.15 Change across a 5-year period in electrical goods

In the *variety shop/furniture* category, there was a correlation between the walk-through frequency and new shop openings, where the rate of new shop openings tended to rise when there were greater numbers of pedestrians. There were new shop launches in 8 out of 12 shops on Higashi-Niomon-Dori St., 10 out of 23 shops on Banshoji-Dori St., and 3 out of 8 shops on Osu Kannon-Dori St. and Shintenchi-Dori St.; however, an increase in the numbers of visits was only seen on Banshoji-Dori St. and Osu Kannon-Dori St.

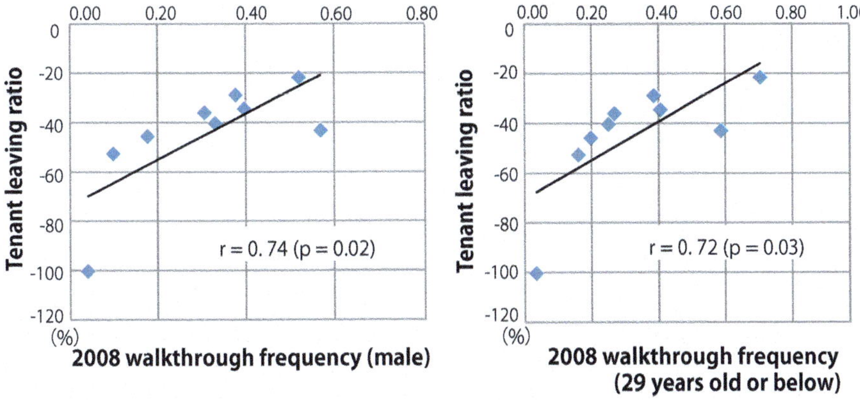

Fig. 2.16 Visit count and launches/closures ratio of shops in the clothing category

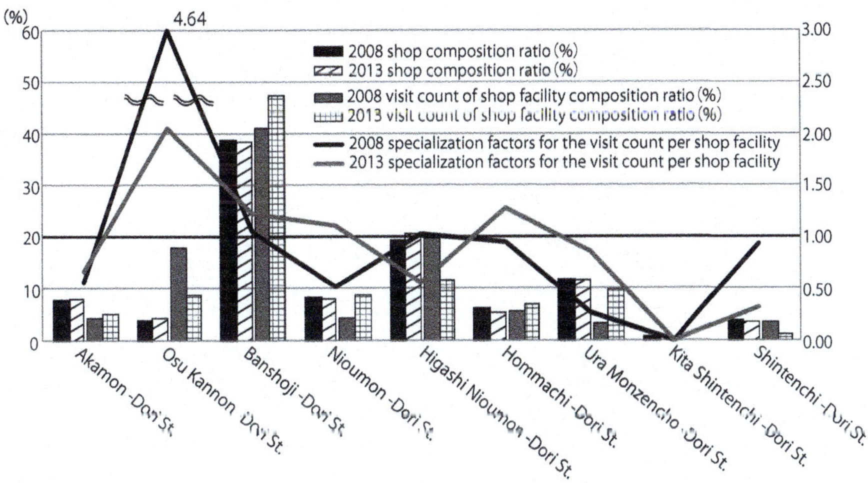

Fig. 2.17 Change across a 5-year period in the clothing category

In the *electrical goods* category, although there were no factors in the shop launches/closing down ratio, there were new shop openings in 5 out of 15 shops on Akamon-Dori St. and 1 out of 4 shops on Shintenchi-Dori St., where a decrease in the number of visits had been identified in 2008, showing an improvement in both visit count and visit count per shop facility. Even on Ura Monzencho-Dori St., there were new shop launches in 1 out of 4 shops, and there was an increase in the visit count.

In the *clothing* category, there was a correlation between the walk-through frequency and the shop closing figures, which can be explained by the fact that the shop closing figures tended to increase according to a lower number of pedestrians. There were shop closures in 11 out of 21 shops on Ura Monzencho-Dori St., 5 out of

11 shops on Hon machi-Dori St., and 6 out of 15 shops on Niomon-Dori St., with an increase in both the visit count and the visit count per shop.

2.5 Conclusion

The findings gained from these analyses are shown below:

1. In 2013, a study was conducted across the nine main streets in the Osu district of Nagoya in Japan, looking at configurations of shops by type, i.e., the product or service on sale in each shop. Compared with an earlier such study in 2008, there had been an increase in the number of shops in the categories of *food and drink/ groceries* and *variety/furniture*, and a decrease in those classified under *clothing*. Further, a change in the configuration of shops by category was confirmed by analyzing the nature of the transitions from one type of commodity to another.
2. Each of the studies included a survey of shop-around behaviors in the target district. The average visit duration/walking distance of visitors to the area was found to have increased over the 5 years, while the number of shops visited had decreased. There was also a decrease in the amount of time spent in individual shops, indicating a change away from planned shop visits and toward walking further distances for longer periods. In terms of walk-through frequency in each zone, there was an increase to the north of Banshoji-Dori St. and a decrease to the south. The shop visit count per person correspondingly increased in the north of the district and decreased in the south.
3. Utilizing the data on changes in shop types within the district, factors driving those transitions from 2008 to 2013 emerged. In the *food and drink/groceries*, *variety/furniture*, and *clothing* categories, differences in shop visit count per person and zone walk-through frequency over the 5-year period were seen to be influenced by the dynamics of shop type change. Additionally, a relationship between the amount of walking around by visitors and the density of shops was confirmed.

References

Arakawa M, Kaneda T (2002) Analyses on redundancy of shop-around behavior in Nagoya CBD. J Architect Plann Environ Eng 67(556):227–233

Hayashi K, Ikemoto M, Kaneda T, Koyama Y, Nakamura J (2013) Actual surveys on visitor's shop-around behaviors and vertical floor use in Akihabara, Tokyo, 2011. AIJ J Technol Design 19(41):315–319

Miyamoto Y, Yuzawa A (2004) Forecast of land use and actual situation of round trip in central Urban District: a case study in Maebashi central Urban District. City Plann Rev. Special issue, papers on city planning 39:661–666

Oiwa Y, Yamada T, Misaka T, Kaneda T (2005) A transition analysis of Shopping District from the view point of visitors' shop around behaviors: a case study of Ohsu District, Nagoya. AIJ J Technol Design 11(22):469–474

Park H, Satoh S (2006) Analysts on the pedestrian rambling activities and spatial structure in downtown: by the follow-up-pedestrian survey using rambling unit. Trans AIJ J Architect Plann 71(605):143–150

Takahashi H, Goto H, Sakuma Y, Saito R, Ishii T (2005) The relationship between the movements of Visitor's walking around and the density of stores on commercial area: a case study on the area around Shimokitazawa-Station. J City Plan Inst Jpn 40:109–109

Takeuchi M, Yoshida T, Kaneda T (2011) A transition analysis of shopping street complex district from the viewpoint of shop-around behaviors: a survey of Osu district, Nagoya. Trans AIJ J Architect Plann 76(660):361–368

Yamada M, Sugiyama S, Tokuono T, Oku T (2009) The change of Osaka Nippombashi electric shopping street: a proposal for revitalization through the change in type of shops. Trans AIJ J Architect Plann 637:611–616. Accessed 30 Mar 2009

Chapter 3
Customer's Spatial Behaviors Inside a Supermarket

Masakazu Takahashi and Masaki Kitazawa

Abstract This chapter examines a small range of movement behavior and discusses the behavior of customers in the supermarkets of Japan as a narrow range of behaviors. First, from the results of existing research on movement behavior, we investigate the direction of research that targets this narrow scope and its issues. Then, we review the social environment in Japan. An overview of aging in Japan, the industrial trends of the retail industry, and those in the target supermarket category are presented. Next, the transaction data from actual stores is analyzed, and the tendency to sell products together is analyzed. Furthermore, to capture the information regarding customer movements in stores, we report the results of investigating customer movements by installing RFID antennas in the store. From the results of analyzing the tendency to sell products together and the results of in-store behavior recording experiments using RFID antennas, we were able to confirm the presence of customers who traveled around the store twice.

3.1 Introduction

According to the latest census, the total population of Japan as of 2017 was 126.71 million. The population over the age of 65 was 35.15 million, accounting for 27.7% of the total population. Furthermore, the number of households was 53.43 million, an increase of 2.8% since 2013. Also, the number of people per household has decreased from 2.46 in 2010 to 2.38, and household sizes continue to shrink. In other words, in Japan, the population has not increased, the proportion of older people has increased, and the labor production population has gradually

M. Takahashi (✉)
Graduate School of Innovation and Technology Management, Yamaguchi University, Yamaguchi, Japan
e-mail: masakazu@yamaguchi-u.ac.jp

M. Kitazawa
Tokyo Institute of Technology, Azbil Corporation, Tokyo, Japan
e-mail: m.kitazawa@trn.dis.titech.ac.jp

T. Kaneda (ed.), *Downtown Dynamics*, Agent-Based Social Systems 16,
https://doi.org/10.1007/978-4-431-54901-7_3

decreased. In such a social environment, it is necessary to redesign various systems such as medical care and pension systems. At the same time, all economic activities require higher efficiency than before. Management points for efficiency need to include not only optimization of product orders such as inventory management, but also labor shifts and cash flows. For retailers, especially those that handle fresh foods, the primary trade area is said to be within 1.5 km from the store location. Since the 2000s, some of the products sold at food supermarkets have overlapped with the products sold at drug stores and convenience stores. The primary trade area of a convenience store is said to be 500 m from the store location. Therefore, competitors have become dense within the trade area.

Furthermore, with the explosive spread of personal information terminals since 2010, various kinds of information have become intermingled with society, regardless of correctness. As a result, it is possible for buyers to obtain information such as sales promotions based on the information gap between sellers and buyers and based on the price gaps between regions, and it is not possible to know where the seller should make use of the characteristics of the current trade area. Traditionally, competition within a trade area has been limited, but B2C (business-to-consumer) companies can also enter the market, resulting in sluggish sales in the retail industry as a whole.

So far, the recommended functions centered on B2C companies have been considered as one of such breakthroughs. However, the recommended function included in B2C is a system that can prompt additional purchases by making a recommendation immediately before product settlement as a result of searching the product database. However, in the retail industry, the purchase result is recorded after the customer's payment, and as a result, promotion beyond the next visit is not achieved. Retailers, especially supermarkets, place shopping baskets on their left sides because most customers are right-handed. For this purpose, the customer flow is designed to be counterclockwise. Product categories that require temperature control and post-purchase processing are fixed on the wall of the sales floor.

For this reason, the products that customers need can be considered as route optimization problems. Under these conditions, it is necessary to project the exact path of the customer based on a full discussion of the customer's in-store trajectory before worrying about the downturn in the retail industry resulting from social problems such as declining birthrates and aging populations. Therefore, in this section, we discuss the behavior of excursions in supermarkets among retailers and capture customer migratory behaviors.

The rest of this chapter is organized as follows; Section 3.2 describes the background and related work on migratory behavior in supermarkets. Section 3.3 briefly explains the data acquisition methodology. Section 3.4 describes a data summary derived from the data acquisition and the proposed model is briefly explained based on the knowledge derived from related work. The proposed model is evaluated and discussed in this section. Section 3.5 discusses the analytical results. In Sect. 3.6, future work and concluding remarks are described.

3.2 Backgrounds and Related Work

In this section, we describe retail stores as exhibiting narrow movement behavior. The target industry is outlined from the Census of Commerce in The Ministry of Economy Trade and Industry (2016). The retail industry and the subcategories of department stores and supermarkets are compared. Data have been gathered every 3 years from 1989 to 2016. Figure 3.1 shows changes in the number of stores and employees. Figure 3.2 shows the changes in annual sales and of the sales floor space.

As Figs. 3.1 and 3.2 show, the number of stores in the retail industry is decreasing. As shown in Fig. 3.2, from the latest census results, although sales floor areas have increased, sales amounts have decreased, so the sales efficiency of the retail industry has decreased. Table 3.1 compares the sales amounts per square meter in the retail industry and for the department stores and supermarkets subcategories. As shown in Table 3.1, not only does the efficiency of the retail industry decline year by year, but the sales efficiency of the sales floor areas of department stores and supermarkets has declined compared to the retail industry, based on data from the Census. Department stores and supermarkets are declining in their sales per unit of sales floor area. Compared to the years 1989 and 2016, the sales efficiency is roughly half. In addition to the decline in sales per unit area, Japan cannot overlook the effects of an aging society.

We now describe the population composition of Japan. Changes in the population per household are summarized. According to the Census in 2016, the number of households was 53.43 million, an increase of 1,453,000 from 2012 and an increase of 2.8%. The number of people per household has decreased from 2.46 in 2012 to 2.38, and household sizes continue to shrink (Statistics Bureau M of IA and C 2016).

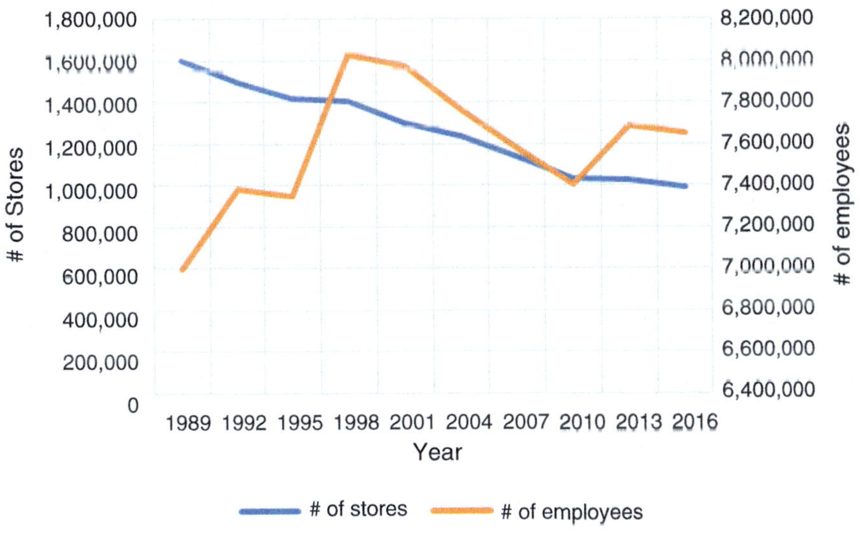

Fig. 3.1 Trends in the number of stores and employees in the retail industry

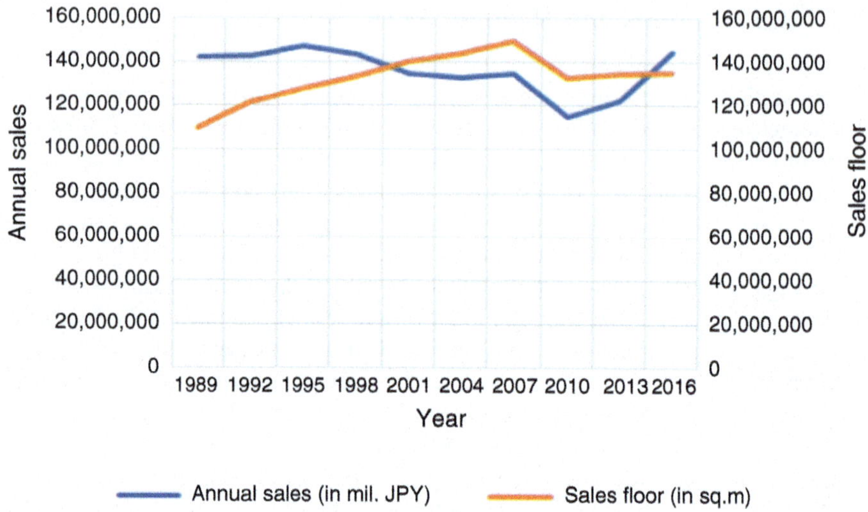

Fig. 3.2 Trends in annual sales and sales floor space for the retail industry

Table 3.1 Comparison of sales amounts per square meter for the retail industry and for the department store and supermarket subcategories

Sales by sqm	Retail industry	Department stores and supermarkets	Differences in %
1989	1.29	1.22	5.7%
1992	1.18	1.08	8.5%
1995	1.15	0.97	15.6%
1998	1.07	0.90	16.5%
2001	0.96	0.78	18.8%
2004	0.92	0.76	18.1%
2007	0.90	0.72	19.8%
2010	0.86	0.62	28.2%
2013	0.91	0.63	30.3%

Table 3.2 Aging ratios (in %) as of October 1, 2017

	Total	Male	Female
Under 15	12.3	12.9	11.7
15–64	60.0	62.3	57.7
Over 65	27.7	24.8	30.6
(65–74)	13.9	13.7	14.2
(Over 75)	13.8	11.1	16.4
Total	100.0	100.0	100.0

Aging in Japan is now summarized. According to the 2018 White Paper on Aging Society, the total population of Japan was 126.71 million as of October 1, 2017. The population older than 65 years was 35.15 million, and its proportion of the total population (aging rate) was 27.7% as shown in Table 3.2. Looking at the population older than 65 years by gender, there were 15.26 million males and 19.89 million females, and the sex ratio (male population for 100 females) was 76.7; the male to female ratio was approximately 3 to 4. In the population aged 65 years and older, the population aged from 65 to 74 years was 17.76 million (8.43 million for men, 9.24 million for women, 91.2 in gender ratio), accounting for 13.9% of the total population; for the population aged 75 and over, the total was 17.48 million (6,840,000 men, 10.65 million women, sex ratio 64.2) for those aged 75 and over. This figure accounts for 13.8% of the total population. The population over 65 years old in Japan was less than 5% of the total population in 1950, exceeded 7% in 1970, and was greater than 14% in 1994. The aging rate has continued to rise and reached 27.7% as of October 1, 2017 (Statistics Bureau M of IA and C 2016; Cabinet Office 2019).

The measures to maintain or improve supermarket sales in such an aging society are described. Pareto optimality, called the 2 to 8 law, with 20% of the top customers comprising 80% of the sales, also applies to the purchasing behavior of supermarket customers (Box GEP and Meyer 1986; Bunkley and Joseph 2018). In addition, the one-to-five rule that acquiring new customers costs five times as much as existing customers suggests that maintaining existing customers is important (Reichheld and Schefter 2000; Reichheld 1990). Therefore, it is reasonable to increase sales by taking sales promotion measures for existing customers and to encourage the purchase of one more item in addition to the original purchase. The Institute of Distribution Economics points out that one of the factors that increases the number of purchases is to increase the length of the customer's in-store flow (The Distribution Economics Institute of Japan 2008).

Also, Tajima et al. (2008) classified three types of marketing research as in-store behavior analysis research methods. (1) The sales audit method based on an analysis of existing data, (2) The method of interviewing customers at the store, and (3) The store test method for actual sales promotion measures. These methods are costly for identifying customer behaviors in stores. Therefore, various methods have been proposed to compensate for the problems of conventional behavior analysis methods in stores. IT technology has been applied to establish an efficient in-store behavior analysis method. Customer purchase history analysis using POS data has been covered by much research (Yada 2006; Yasuda et.al. 2014; Namba et al. 2016).

Retail stores conduct various sales promotions to improve sales, such as POP (Point Of Purchase) advertisements, unique displays, and in-store decorations. To predict the effectiveness of sales promotions, it is necessary to analyze customer behavior.

For example, suppose an in-store experiment is performed; there is not only a very high cost, but there is a risk of customers leaving the store if we do not consider their approval of the experiment. On the other hand, the use of mathematical analyses such as data mining is useful as one of the methods for understanding, but these also produce analytical results that are complicated to understand. Therefore, if an agent-based simulation (ABM) is employed for such analytics, in-store experiments can be performed virtually. Compared with actual in-store experiments, it can be done quickly and at low cost. Furthermore, it is possible to consider the influence of the store environment on customers. Therefore, in-store movements

Table 3.3 Study of in-store movements

Purpose	Purchasing behavior	Migrating behavior	Remarks	Reference #
Proposal for a customer behavior model	×	×	Model proposal only	Toyoshima et al. (2005)
Proposal for a sales floor layout	×	×	Comparison of simulation results	Tajima et al. (2008)
Strategic evaluation of digital signage	×	×	Comparison of simulation results	Hayashi and Cho (2010)
Proposal for a sales floor layout	POS	Interview	Agent-based in-store simulator (ABISS)	Terano et al. (2009), Kishimoto et al. (2009)
Proposal for a customer behavior model	POS	In-store experiment	Agent-based in-store simulator (ABISS) 2	Kitazawa et al. (2010)

using ABM are considered. Hayashi et al. obtained customer movement data using RFID (Radio Frequency IDentifier) data and analyzed customer movement routes in a store (Hayashi and Cho 2010). Tajima modeled the customer as an agent (Tajima et al. 2008). There is also a method using agent-based simulation (ABS). Kitazawa et al. developed an agent-based in-store simulator (ABISS) to search for sales promotion measures to increase the length of customer stays in stores (Kitazawa et al. 2010). Table 3.3 shows the study on in-store migration.

In-store movements using actual transaction data and customer walking paths are led by simulations using ABM by Kishimoto et al. (2009). In the future, movement behavior can be identified technically using the unique MAC address installed in the smartphone Wi-Fi function. However, due to the restrictions of personal information protection law in Japan, it is only technically possible at this stage. Since it is possible to derive the movement routes for all customers, it is expected that the legal system will operate flexibly in the future.

3.3 Data Collection Configuration

The item layouts and the antenna locations in a store are shown in Fig. 3.3. The primary customer flow is designed for traveling counterclockwise from the windbreak room shown in the lower center, product selections, and payment. To simulate store movements in a supermarket, the customer trajectories were fitted with RFID antennas on four fixtures in the store. (1) Entrance/Vegetables, (2) Fish, (3) Meat, and (4) Tofu/Beverage. RFID tags were attached to the sides of the shopping baskets. When a basket with a tag passes in front of an antenna, the antenna number, tag number, date, and time are recorded. RFID tags were attached to both sides of 100 shopping carts among a total of 300 shopping carts. Data were collected within business hours from 9:00 to 21:00 on July 19th to 26th, 2015.

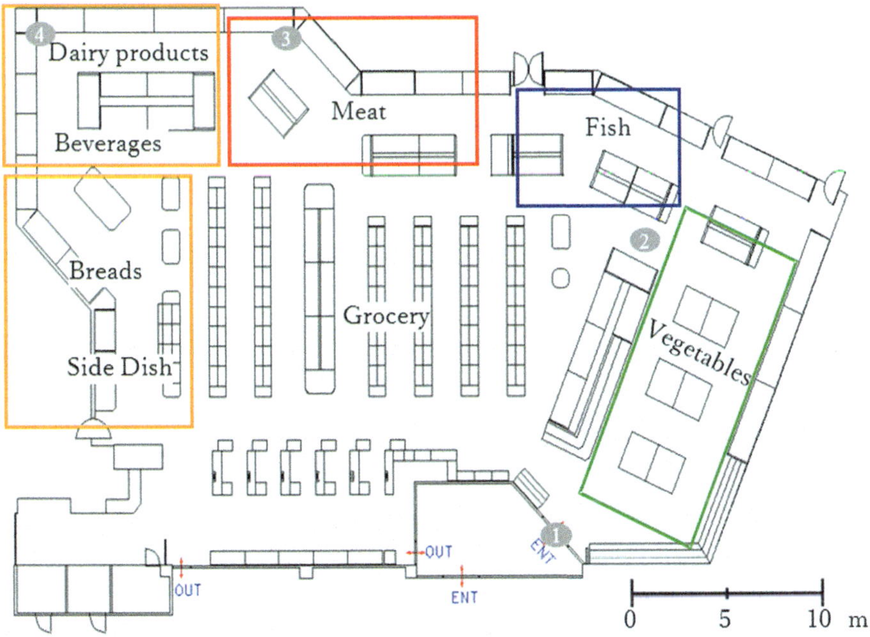

Fig. 3.3 Category layout and antenna locations in the target store

3.4 Analytical Results

The total number of samples used in the analysis is 432. Table 3.4 shows the number of times that the RFID tags passed each antenna. Table 3.5 also indicates the number of customers who made one round with their round patterns. Based on the list in Table 3.4, the IDs of the RFID tags and the recorded antenna numbers and times were integrated to obtain the movement route for each transaction. As a result of simple aggregation, 38 movement patterns were generated from the four antennas. The flow pattern for July 22 is shown in Tables 3.6 and 3.7. Outlines of in-store laps were 360 (83.3%) for one lap, 66 (15.3%) for two laps, and were 6 (1.4%) for three

Table 3.4 Customer cycle pattern for one round

Location	#
Antenna #1	467
Antenna #2	332
Antenna #3	256
Antenna #4	111

Table 3.5 Customer cycle pattern for one round

Pattern	123	1	12	1234	13	134	124	14	Total
#	87	84	81	46	28	14	12	8	360

Table 3.6 Customer cycle pattern for two rounds

Pattern	12,343	121	132	1231	
#	7	6	6	5	
Pattern	12,321	1232	12,342	131	
#	5	5	4	4	
Pattern	142	12,323	1212	12,341	
#	3	3	2	2	
Pattern	1213	1,234,234	123,432	12,412	
#	2	2	1	1	
Pattern	12,313	123,212	123,412	1,234,323	
#	1	1	1	1	
Pattern	1243	1323	13,423	143	Total
#	1	1	1	1	66

Table 3.7 Customer cycle pattern for three rounds and more

Pattern	1,212,342	12,143	12,323,212	
#	1	1	1	
Pattern	121,242	1,232,321	123,232,432	Total
#	1	1	1	6

Table 3.8 Association rules

Item A	Item B*i*	Supp (A → B*i*)	Conf (A → B*i*)	Lift (A → B*i*)
Sausage	Meat (pork)	0.062	0.438	1.985
	Milk	0.029	0.202	1.518
	Fish paste	0.037	0.260	1.847
	Vegetables	0.073	0.514	1.592
	Egg	0.039	0.277	1.666
	Noodle	0.016	0.112	1.729

laps or more. Based on the data collected during the data collection period, a combined purchase rule for the meat category in the antenna installation location was generated.

Based on the data gathered during the period, the combined purchase rules for the meat category at the third antenna location were generated. The data used in this chapter have a JAN (Japan Article Number) code assigned to the individual product ID of the transaction unit. Using this code, sales volumes and sales prices are obtained. The data hierarchy has four layers for product management: Department, Line, Class, and JAN.

This time, the in-store movements obtained by extracting the purchase rules in the Line layer are estimated. The purchase rule uses three indicators: Support, Confidence, and Lift.

The three indices are defined by the following equation (Agrawal et al. 1993, 1996; Borgelt and Kruse 2002). According to the calculation method shown in Formula (3.1) to Formula (3.3), the rules for the meat division are shown in Table 3.8.

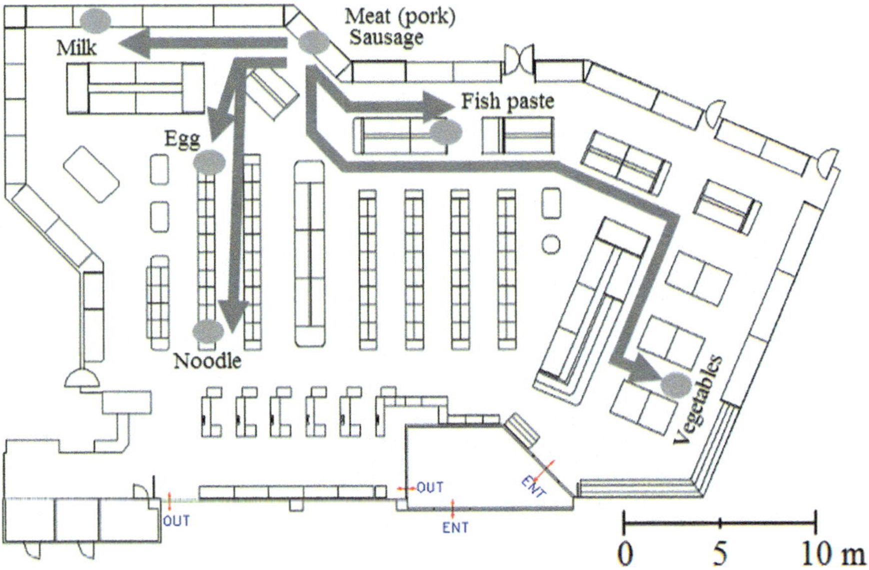

Fig. 3.4 Rules and item locations

Support indicates the proportion of transactions in which items A and B were purchased at the same time for all transactions. Define X is as all transactions.

$$\text{Support}_{\text{A}\to\text{B}} = |\text{A} \cap \text{B}|/X \tag{3.1}$$

Confidence refers to the proportion of transactions where B is purchased at the same time, that is, the conditional probability, of the transactions where product A is purchased.

$$\text{Confidence}_{\text{A}\to\text{B}} = |\text{A} \cap \text{B}|/|\text{A}|. \tag{3.2}$$

Lift is the value divided by the purchase probability when A and B are assumed to be independent. It is an indicator of how much the probability is increased/lifted by the purchase of A.

$$\text{Lift}_{\text{A}\to\text{B}} = \left(\frac{|\text{A} \cap \text{B}|}{|X|}\right) / \left(\frac{|\text{B}|}{|X|}\right) \tag{3.3}$$

Based on Formula (3.1) to Formula (3.3), the rules for food categories were investigated close to antenna 3. Table 3.8 shows the results of products with high lift values in the vicinity of antenna three and their association rules.

Figure 3.4 shows the positions of the item categories located near antenna 3. From the in-store experiment using RFID, the movement behavior model confirmed the existence of customers who passed twice around the store.

3.5 Discussion

Japan's aging rate is very high at approximately 27%. Sarcopenia is the decrease in skeletal muscle mass and skeletal muscle strength with age. With aging, this leads to a significant decrease that leads to progressive decreases in skeletal muscle mass and physical strength and function (Phillips 2015; Ryall et al. 2008; Marcell 2003; Doherty 2003). It cannot be denied that decreases in walking speed and shortened strides due to decreases in muscular strength affect consumption behaviors of the elderly. Layouts that are comfortable for everyone will become increasingly important in the future.

Yamada et al. (2013) focused on the existence of people who have difficulty shopping and studied a method for predicting the occurrence of people with shopping difficulties, considering the space demand for retail services. However, this is an effort in the entire block space but considers the decline in mobility associated with aging. In the future, it will become increasingly important to design stores that provide high mobility for the elderly, considering sarcopenia.

3.6 Concluding Remarks

In this chapter, to understand the properties of store movements, we installed antennas in a store and investigated customer movements. From the results of the analytics, the position of the item category installed near antenna 3 was extracted. As a result, from the in-store experiment using RFID, the movement behavior model confirmed the existence of customers who go around the store twice. To maintain or improve sales, it is necessary to increase customer purchases by encouraging customers to purchase. In this chapter, we employed association rules to estimate customer movements based on the rules for buying items together.

Acknowledgments This work was ostensibly supported by the collaborative research fund from a domestic retail store. The authors gratefully acknowledge the work of past and present members of our laboratory for their helpful discussions and comments on the manuscript.

References

Agrawal R, Imieliński T, Swami A et al (1993) Mining association rules between sets of items in large databases. In: Proceedings of the 1993 ACM SIGMOD international conference on management of data—SIGMOD '93. ACM, New York, pp 207–216

Agrawal R, Mannila H, Srikant R et al (1996) Fast discovery of association rules. Adv Knowl Discov Data Min 12:307–328

Borgelt C, Kruse R (2002) Induction of association rules: a priori implementation. In: Härdle W, Rönz B (eds) Compstat. Physica-Verlag HD, Heidelberg, pp 395–400

Box GEP, Meyer RD (1986) An analysis for unreplicated fractional factorials. Technometrics 28:11–18. https://doi.org/10.2307/1269599

Bunkley N, Joseph J (2018) 103, Pioneer in Quality Control, Dies. The New York Times. https://www.nytimes.com/2008/03/03/business/03juran.html. Accessed 11 Sep 2019

Cabinet Office (2019) Annual report on the Ageing Society. https://www8.cao.go.jp/kourei/english/annualreport/index-wh.html. Accessed 11 Sep 2019

Doherty TJ (2003) Invited review: aging and sarcopenia. J Appl Physiol 95:1717–1727. https://doi.org/10.1152/japplphysiol.00347.2003

Hayashi H, Cho K (2010) Evaluating in-store digital signage by agent-based simulation. In: 3rd World Congress on Social Simulation (WCSS2010). Kassel, Germany

Kishimoto A, Takahashi T, Takahashi M et al (2009) An agent simulator for analyzing consumer behaviors and sales promotion in a retail store. IEICE Tech Rep 108:87–92

Kitazawa M, Takahashi M, Yamada T et al (2010) How do customers move in a supermarket?: analysis by real observation and agent simulation. Inst Electr Eng Jpn 2010:17–20

Marcell TJ (2003) Sarcopenia: causes, consequences, and preventions. J Gerontol A Biol Sci Med Sci 58:M911–M916. https://doi.org/10.1093/gerona/58.10.m911

Namba Y, Yoshida EJ, Tokunaga K, Haraguchi T (2016) Featured articles. AI services and platforms: a practical approach to increasing business sophistication. Hitachi Rev 65

Phillips SM (2015) Nutritional supplements in support of resistance exercise to counter age-related sarcopenia. Adv Nutr 6:452–460. https://doi.org/10.3945/an.115.008367

Reichheld F (1990) Zero defections-quality comes to services. Harv Bus Rev 105–111 https://hbr.org/1990/09/zero-defections-quality-comes-to-services. Accessed 11 Sep 2019

Reichheld F, Schefter P (2000) E-Loyalty: your secret weapon on the web. Harv Bus Rev 78. https://hbr.org/2000/07/e-loyalty-your-secret-weapon-on-the-web. Accessed 11 Sep 2019

Ryall JG, Schertzer JD, Lynch GS (2008) Cellular and molecular mechanisms underlying age-related skeletal muscle wasting and weakness. Biogerontology 9:213–228. https://doi.org/10.1007/s10522-008-9131-0

Statistics Bureau M of IA and C (2016) Outline of the 2015 population census of Japan. In: Statistics Bureau. https://www.stat.go.jp/english/data/kokusei/index.html. Accessed 11 Sep 2019

Tajima T, Abe T, Nambo H, Kimura H (2008) Simulator of consumption behavior with the customer's attributes and merchandise interrelationships. In: The 22nd Annual Conference of the Japanese Society for Artificial Intelligence. The Japanese Society for Artificial Intelligence, Asahikawa, Hokkaido, p 3F4–6

Terano T, Kishimoto A, Takahashi T et al (2009) Agent-based in-store simulator for analyzing customer behaviors in a super-market. Springer, Berlin, pp 244–251

The Distribution Economics Institute of Japan (2008) ISM: in-store merchandising. Nikkei, Tokyo

The Ministry of Economy Trade and Industry (2016) Census of Commerce. In: Ministry of economy, trade and industry. https://www.meti.go.jp/english/statistics/tyo/syougyo/index.html. Accessed 11 Sep 2019

Toyoshima I, Koiso T, Yoshida T et al (2005) One day customer-go-around model in a retail store, based on ubiquitous data. IPSJ SIG Notes ICS 2005:115–120

Yada K (2006) Consumer behavior analysis by graph mining technique. New Math Nat Comput 2:59–68

Yamada T, Kii M, Doi K, Itami E (2013) Estimation of shopping refugees considering spatial supply-demand balance of retail service. J Japan Soc Civ Eng Ser D3 (Infrastruct Plan Manag) 69:I_91–I_99. https://doi.org/10.2208/jscejipm.69.I_91

Yasuda M, Namba Y, Yoshida EJ (2014) New business trends created by big data utilization nature and purpose of big data. Hitachi Rev 63

Part II
Spatial Distribution of Prosperity and Visibility

Chapter 4
Analysis of the Correlation Between Underground Spatial Configurations and Pedestrian Flows Using Space Syntax Measures: A Case Study of the Sakae District Underground Mall Complex

Akira Ota, Rikako Mizuno, Gadea Uriel Garrido, and Toshiyuki Kaneda

Abstract Since the early stages of Natural Movement research in the 1990s, Space Syntax (SS) measures were used as a factor in the analysis of pedestrian numbers. Because an underground mall is usually higher in geographical closure than the ground, it can be a good specimen with which to discuss the effectiveness of the SS measure in the factor analysis of pedestrian spatial distributions.

In our research, a pedestrian gate count survey was conducted of an underground mall complex of Sakae District in Nagoya, and a factor analysis of the pedestrian numbers was carried out, assuming the inside of the underground mall complex to be a closed system. Therefore, explanatory factors were extracted in the order of visual step depth from the station ticket gate, the visual step depth from the adjoining building via an underground passage, and the counts of directly visible shop tenants. These three variables supported the effectiveness of the SS measure as a factor describing pedestrian numbers.

A. Ota (✉)
Keio University, Yokohama, Kanagawa, Japan

Tokyu Land Corporation, Tokyo, Japan
e-mail: akira_ota@keio.jp

R. Mizuno
Toyota City Office, Toyota City, Aichi Prefecture, Japan

G. Uriel Garrido
ONOCOM Co. Ltd, Tokyo, Japan

T. Kaneda
Graduate School of Engineering, Nagoya Institute of Technology, Gokiso, Showa, Nagoya, Japan
e-mail: kaneda@nitech.ac.jp

T. Kaneda (ed.), *Downtown Dynamics*, Agent-Based Social Systems 16,
https://doi.org/10.1007/978-4-431-54901-7_4

4.1 Introduction

4.1.1 Research Background and Objectives

The planning and design of underground shopping malls in Japan are important toward not only improving general pedestrian flows but also optimizing passenger transfers between stations. Such locations are attractive because of their atmosphere of hustle and bustle created by their configurations and the flow of commuters, shoppers, and visitors. One well-known piece of research that studied such crowdedness from the viewpoint of spatial configurations was an analysis incorporating Space Syntax theory (SS theory) by Hillier et al. (1987) of the University College London Group. SS theory is an approach that quantifies the conditions of an urban form; it has been reported by Hillier et al. (1993) that SS measures are effective when used to identify key factors in the creation of crowdedness as will be described later. However, the intensity of the factors expressed by each measure and their validity are still up for debate, and particularly considering the dearth of reports on the application of SS measures to underground malls, it is meaningful to accumulate case studies for analysis. As a feature of this research, shop proximity is considered to be important to the pedestrian flow of an underground mall; therefore, in addition to the SS measures used in prior research analysis, a newly developed measure that considers shop proximity in underground malls was also used.

To explore factors affecting pedestrian distribution within the underground mall complex in Sakae District, this work of research measured pedestrian numbers in the complex and conducted multiple regression analyses using SS measures based on visibility graph analysis and underground shop proximities.

4.1.2 Prior Research on the Factor Analysis of Pedestrians Applying SS Measures

Prior to our research, we examined earlier domestic reports published by the Architectural Institute of Japan and the City Planning Institute of Japan and overseas reports presented at the International Space Syntax Symposium, and the International Seminar on Urban Forum among others. These reports concerned the factor analysis of pedestrians by applying SS measures.

In the 1990s, the correlation between pedestrians and SS measures was reported by Hillier et al. (1993) overseas and by Hanazato et al. (1991) in Japan. As for factor analyses of pedestrian numbers, an overseas report by Desyllas et al. (2003) presented an analysis wherein an SS measure (i.e., "visible areas") was included as a factor in addition to the distance from the nearest station, intensity of land use, and pedestrian lane width. Later, Özer and Kubat (2007) reported a factor analysis of pedestrians in Istanbul and concluded that the SS measures were secondary factors. In Japan, Araya et al. (2005) analyzed pedestrian numbers using both an SS measure—street network centrality—and the distance to the nearest station; here, the SS measure was found to be the primary factor and the distance the secondary factor. Afterward, Ota et al. (2008, 2015) reported that the factor analysis of pedestrian numbers, applying SS measures, was effective even when comparing different points in time. The analysis of a multilevel case was reported by Afroza et al. (2007) overseas, while in Japan, Ueno et al. (2009) analyzed the multilevel facilities of a station and reported that SS measures were primary factors.

One of these reports addressed the underground space based in the same area in which this work of research is based. Okamoto et al. (2013) analyzed the correlations between SS measures and pedestrian numbers in the Nagoya Station underground mall complex. They used SS measures to analyze pedestrian numbers and found the distance from the nearest station or ticket gate to be a primary factor and SS measures as secondary factors; however, SS measures were adopted for the statistical tests. They clarified that in the Nagoya Station underground closed space, a combination of the "shortest distance from the nearest ticket gate" and the "centrality of the underground space" influenced pedestrian flows and crowding.

From previous reports' factor rankings, in general, the distance from the nearest station (ticket gate) tended to be ranked higher; however, some reports do give SS measures a higher ranking. In addition, in all reports, SS measures were incorporated into the statistical tests; this suggests the possibility of future pedestrian factor analyses including SS measures as factors.

4.1.3 Structure of the Research

Section 4.1 describes the background, objectives, and structure of this research.

Section 4.2 introduces Sakae District and the location of the case study and describes the details of the pedestrian count surveys.

Section 4.3 calculates the candidate explanatory variables for analysis in this work of research. It outlines the various SS measures and a measure concerning shop proximity in underground malls, and then, the calculation methods are clarified.

Section 4.4 describes the correlation between the candidate explanatory variables and pedestrians, followed by regression analysis using the stepwise method. The analysis results are then discussed.

Section 4.5 presents the conclusion and tasks to be tackled in the future.

4.2 Case Study: Sakae District

4.2.1 Outline of Sakae District

Sakae District is a flourishing business area within Nagoya, one of the leading cities in Japan. It has stations serving three lines: Subway Higashiyama Line, Subway

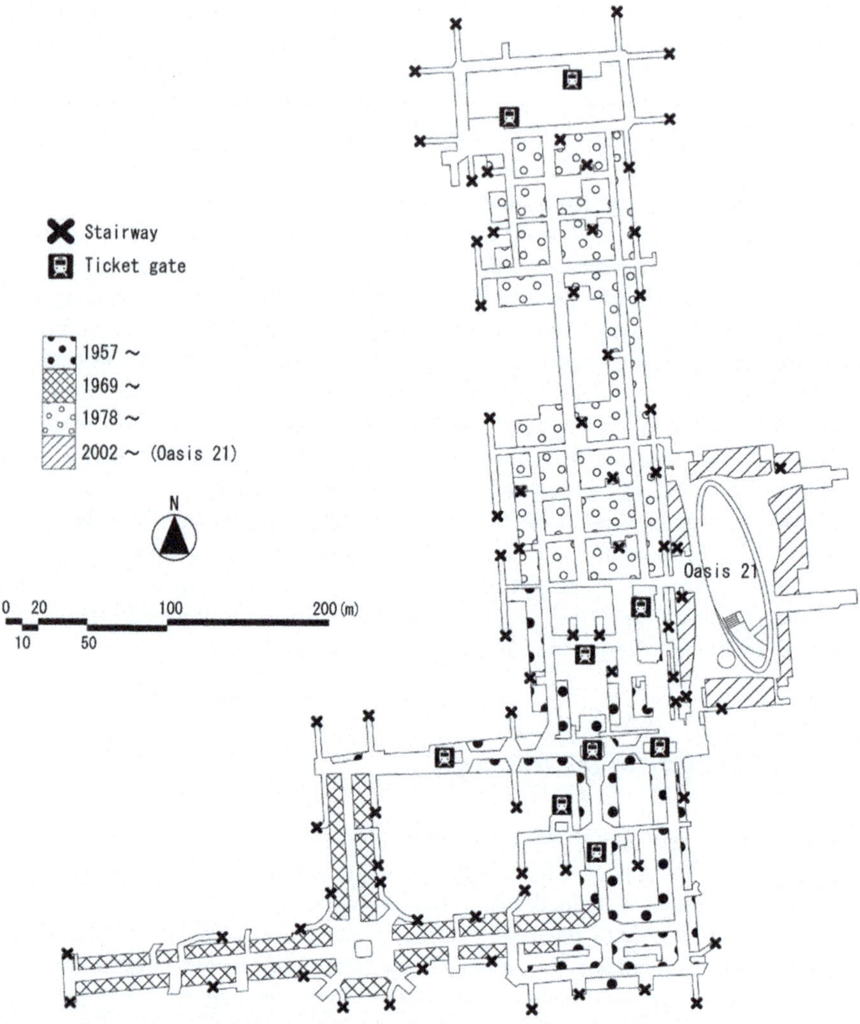

Fig. 4.1 Sakae District underground complex

Meijo Line, and the Nagoya Railroad Seto Line. In 2014, Sakae Station had a daily passenger throughput of approximately 111,000; Hisaya-odori Station had one of approximately 23,000; and Sakaemachi Station had one of nearly 20,000. Presently, the underground complexes of Nagoya City, covering approximately 166,000 m^2, are ranked third largest after Tokyo and Osaka. They are split into two nearly equal areas, Nagoya Station and Sakae District, roughly 83,000 m^2 each. Unlike the labyrinthine Nagoya Station, the Sakae District underground mall complex (Sakae District underground complex) surveyed in this research has a relatively simple structure, extends north–south under the 100-meter-long Hisaya-odori Street, and extends west under Hirokoji Street, a major east–west road (Fig. 4.1). The surveyed area has a total of 74 stairways, connecting the ground level with the underground sections, and nine subway ticket gates, serving the three train lines.

4.2.2 Survey of Pedestrian Numbers in the Sakae District Underground Complex

In September 2016, we conducted a simplified gate count survey in Sakae District (aboveground and underground). Figure 4.2 shows the survey spots and results. The time slot chosen was between 13:30 and 15:30 on a weekday as it was at this hour usually that many strolling pedestrians could be observed. Each of the 30 spots aboveground and 55 spots underground was surveyed for a 5-min period, and the sum of the cross-sectional pedestrian numbers, in both directions, was collected. In the aboveground survey, cyclists were not included in the count.

An examination of the pedestrian distribution showed that the Higashiyama Line and southward area had higher pedestrian numbers when compared with the center of the underground space. There are three possible reasons for this result: (1) the ticket gates extended over two stations; (2) adjacent facilities converged on the south side; and (3) Crystal Square was a popular meeting spot. The pedestrian numbers were found to decrease at spots in a passage with almost no facing shops.

4.3 Candidate Explanatory Variables Including SS Theory Measures

4.3.1 SS Theory and Proposed Measures

SS theory is a general term for a series of spatial analysis techniques proposed by Hillier et al. of University College London in the early 1980s. This work of research used SS measures that were based on the theory's visibility graph analysis.

As for visibility graph analysis, the surveyed area was divided into a grid, and attention was paid to the links between the central points of the sectors so as to define each quantitative measure, among which the following four measures were studied: Connectivity (CNT), Visual Step Depth (VSD), the shortest distance—Metric Shortest Path length (MSP)—and Integration Value (IV). Each measure is outlined below, and Fig. 4.3 shows an image of each SS measure.

CNT is the total number of grid sectors that are directly visible from a sector center; a higher value indicates a broader visible area from that point, and that grid section is visible from many directions.

VSD is the number of perspective lines (depth) and represents the least number of steps between two specified sectors; a step is a visible direction change that is made when moving from one sector to another. When a specific sector is assigned as the

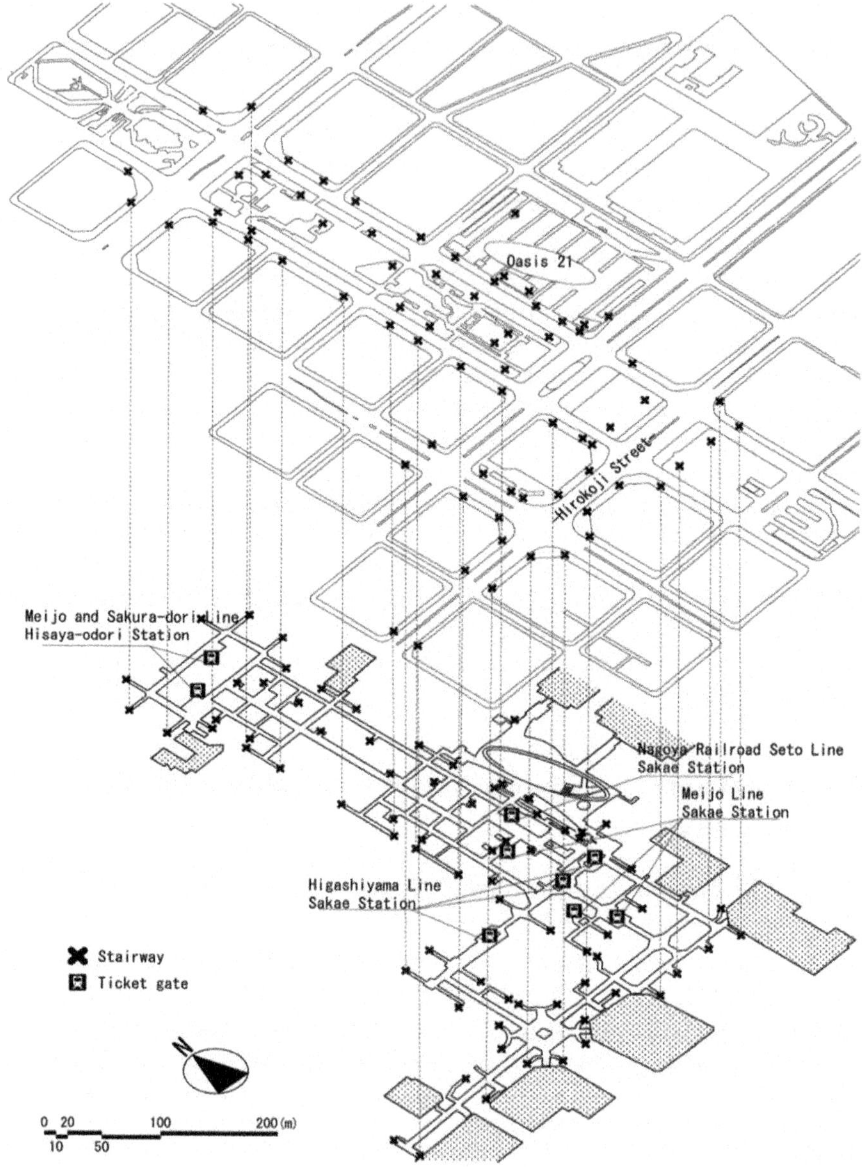

Fig. 4.2 Map of the Sakae District underground complex

zero step, any sectors visible from there are assumed to be one step. A smaller value indicates movement that is visually easy from a sector. In VSD, the concept of distance is abstracted.

MSP is the measure of the Euclidean distance between vertices and is unusual among SS theory measures, a majority of which are dependent on spatial topological

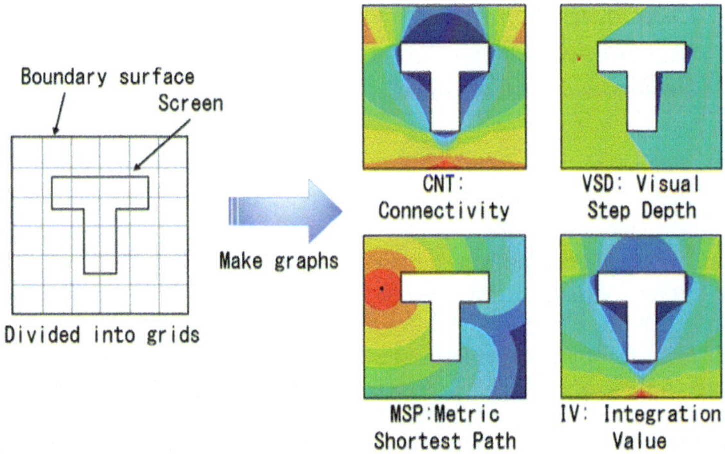

Fig. 4.3 Images of space syntax measures

distances. A smaller value shows the relevant grid sector as being located closer to another specific sector.

IV is the reciprocal of the value obtained by standardizing the average depth. To find IV, the Relative Asymmetry (RA) needs to be found first. A topological distance from one point on a graph to another is regarded as depth, and the average of the depths from one point to all other points is the mean depth (MD), and from this value, the RA value is found (Expression 4.1). Since the RA value is dependent on the vertex to be analyzed, by standardizing the RA value to find the Real Relative Asymmetry (RRA), as shown in Expression (4.3), it becomes possible to compare spaces with differing numbers of vertices. Moreover, to make it easier to understand, a reciprocal value is found, which is the IV (Expression 4.4). The IV has depth as a denominator; therefore, a higher value indicates that the relevant grid has a higher degree of centrality within the whole closed area.

$$\mathrm{RA} = \frac{2(\mathrm{MD} - 1)}{k - 2} \tag{4.1}$$

$$D_k = \frac{2\left[k\left\{\log_2\left(\frac{k+2}{3}\right) - 1\right\} + 1\right]}{(k - 1)(k - 2)} \tag{4.2}$$

$$\mathrm{RRA} = \frac{\mathrm{RA}}{D_k} = \frac{(\mathrm{MD} - 1)(k - 1)}{\left[k\left\{\log_2\left(\frac{k+2}{3}\right) - 1\right\} + 1\right]} \tag{4.3}$$

$$\mathrm{IV} = \frac{1}{\mathrm{RRA}} \tag{4.4}$$

4.3.2 Candidate Explanatory Variables and Spatial Distribution for Spatial Analysis of the Sakae District Underground Complex

In this research study and analysis, the areas covered by the four underground malls of Sakae District were treated as a single area (i.e., one closed case). Depth Map was used as the analysis software for calculating each measure's value in the visibility graph analysis, and grid points were set at 1-m intervals. Where a high or wide screen (of 2 m or more) existed, it was assumed that passages beyond the screen were not visible.

In the analysis of the Sakae District underground complex, the following pairs of candidate explanatory variables were prepared: (1) Visual Step Depth to Station (VSDS) and Metric Shortest Path to Station (MSPS) as proximity measures to the nearest ticket gate; and (2) Visual Step Depth to Adjacent facilities (VSDA) and Metric Shortest Path to Adjacent facilities (MSPA) as proximity measures to

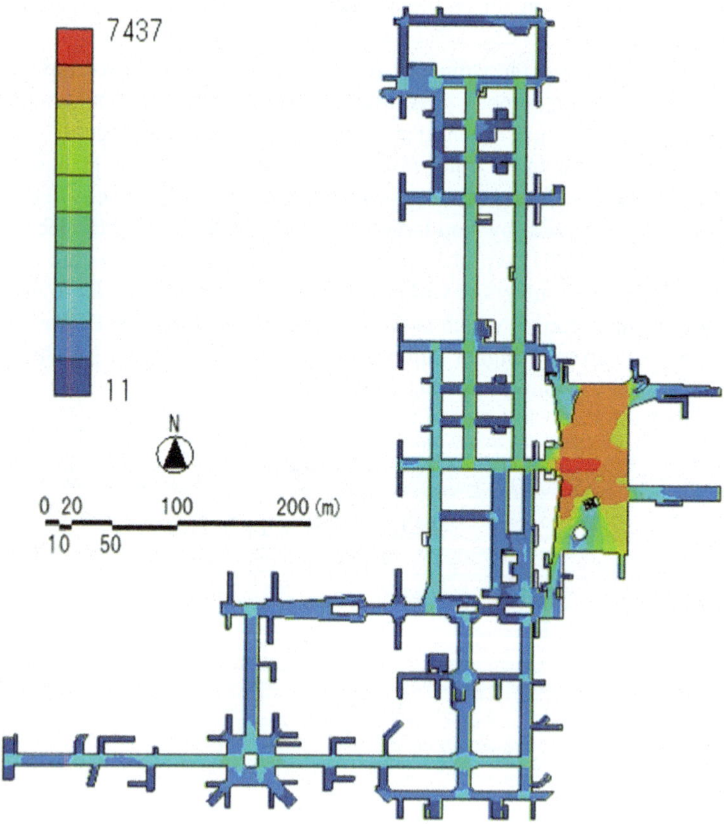

Fig. 4.4 Spatial distribution of connectivity measure values (CNT)

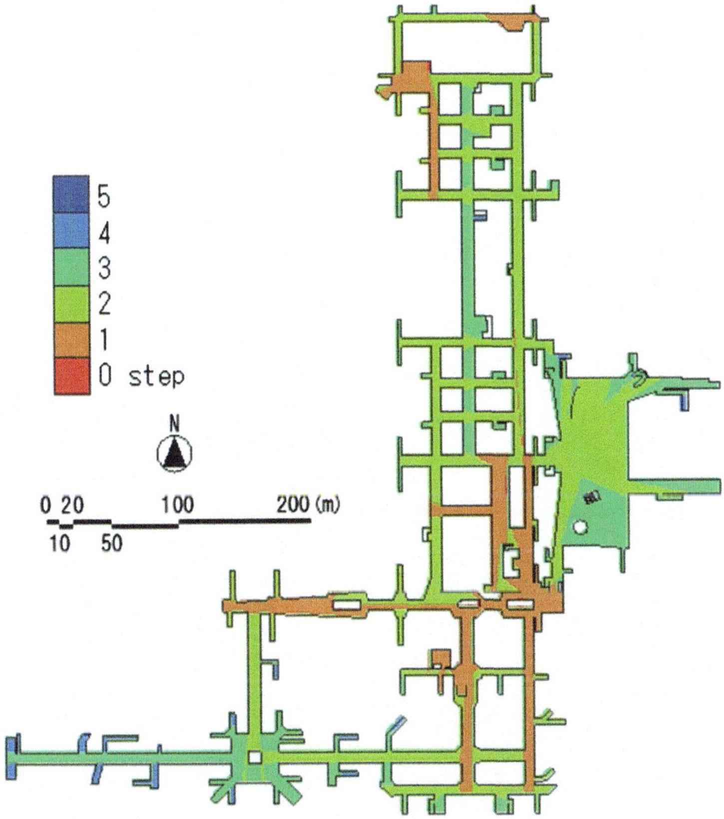

Fig. 4.5 Spatial distribution of visibility depth step to the nearest station ticket gate (VSDS)

adjacent facilities. For the Integration Values (IV), a Global Integration Value (GIV) using an average depth of the whole underground area, and a Local Integration Value (LIV3) with calculations stopping at Depth 3 were prepared.

The spatial distribution of each measure was examined. CNT measure values increased at the Oasis 21 section with its broad visible area, and longer passages showed slightly higher values (Fig. 4.4). The spatial distributions of VSDS and MSPS showed lower values around the centers of any areas where the ticket gates converged (Figs. 4.5 and 4.6). For the distributions VSDA and MSPA, it is noticeable that values decreased at the south side where adjacent facilities were concentrated (Figs. 4.7 and 4.8). From the spatial distributions of GIV and LIV3, it was found that a spot with a higher centrality had a higher value, with values around Oasis 21 found to be particularly high (Figs. 4.9 and 4.10).

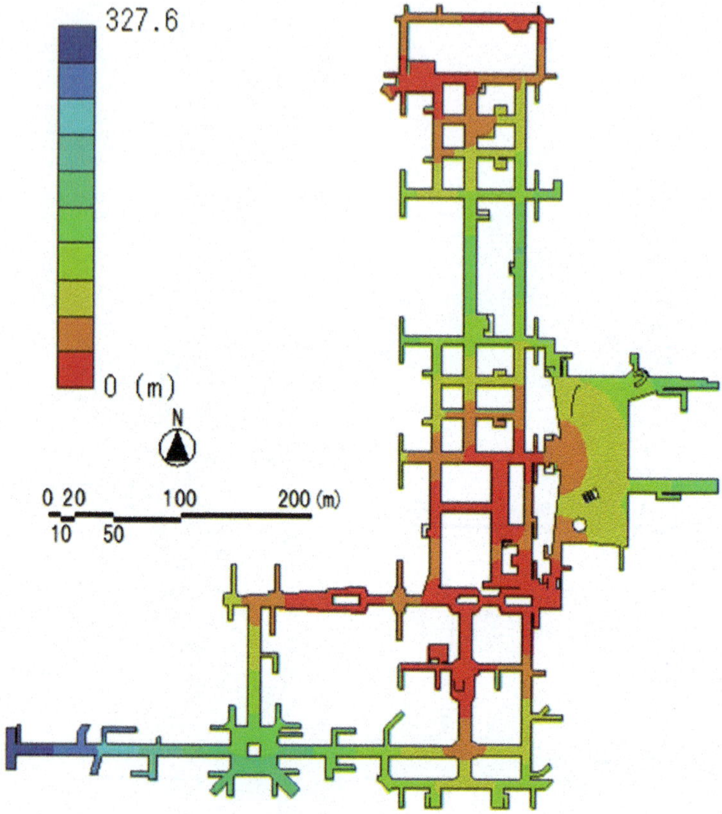

Fig. 4.6 Spatial distribution of the shortest distance to the nearest station ticket gate (MSPS)

4.3.3 Candidate Explanatory Variables for Shop Proximity and Spatial Distribution in the Sakae District Underground Complex

As variables related to shop proximities in the underground complex, this research study developed and used the Tenant Count Depth (TCD) and the Tenant Count Metric (TCM). TCD refers to the number of visible shops in the underground complex that can be visually confirmed (Depth 1) from a pedestrian number survey spot. A higher TCD value indicates a higher number of shops that are easy to locate visually. Figure 4.11 shows the TCD values for the Sakae District underground complex. TCM refers to the number of shops within a 50-m radius from a pedestrian number survey spot. A higher TCM value indicates a higher number of shops at a

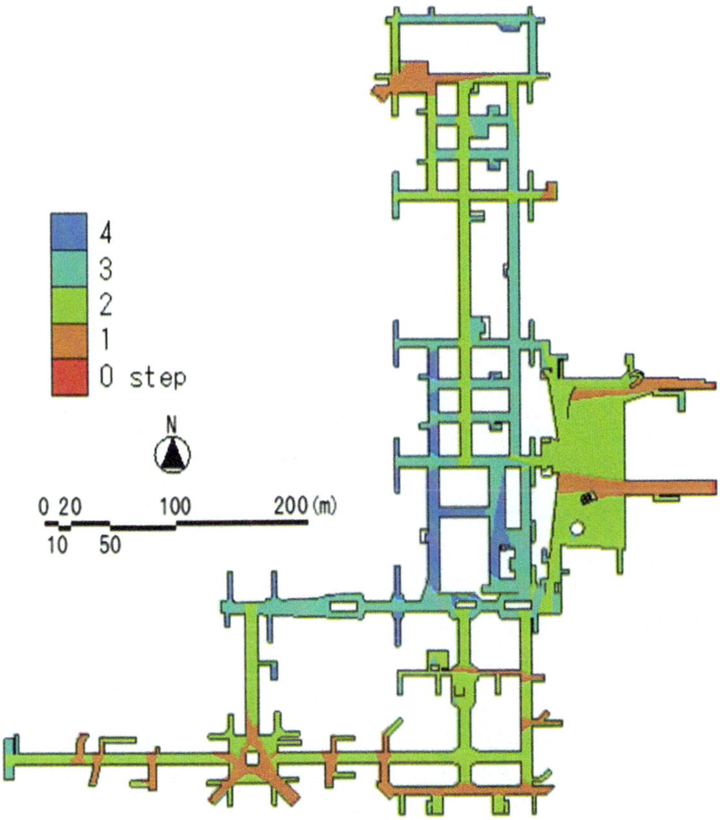

Fig. 4.7 Spatial distribution of the visibility depth step to adjacent facilities (VSDA)

close distance. Figure 4.12 shows the TCM values for the Sakae District underground complex.

Figures 4.11 and 4.12 enable us to confirm that, in the Sakae District underground complex, the TCD values were high in long passages. This is the same tendency as shown by CNT; therefore, it is possible to consider that the values were affected by the areas visible. The TCM values were high at the area from the central part to the south side; it is worth considering that areas with high centralities tended to show relatively high TCM values.

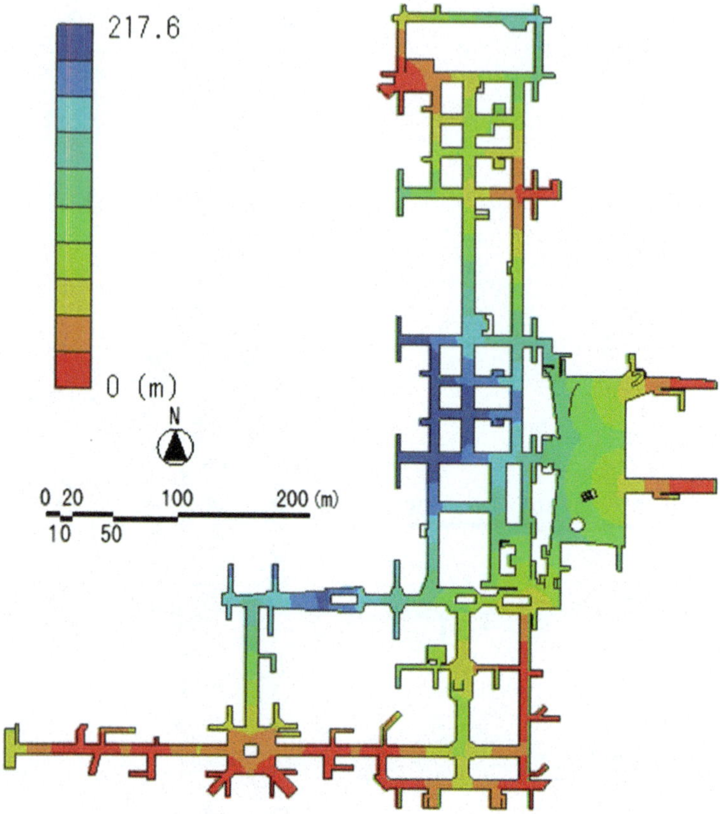

Fig. 4.8 Spatial distribution of the shortest distance to adjacent facilities (MSPA)

4.4 Analysis of the Relationships Between Underground Spatial Configurations and Pedestrian Flows Using Space Syntax Measures

4.4.1 Correlation Between the Pedestrian Numbers and Candidate Factor Variables

The statistical analysis software JMP was used to create a correlation matrix (Table 4.1) of the number of pedestrians who were surveyed at the 55 spots in the underground complex and candidate factor variables.

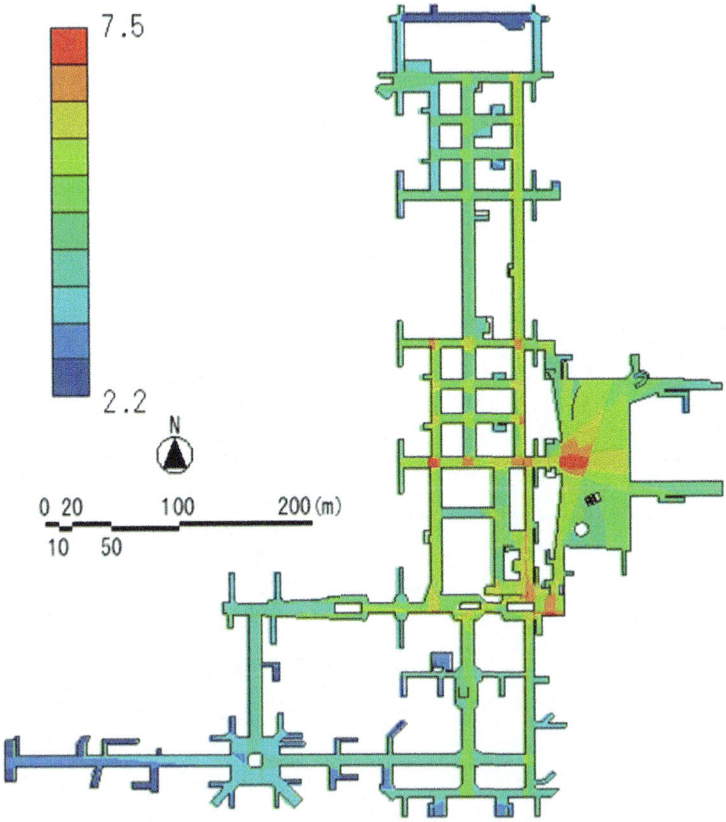

Fig. 4.9 Spatial distribution of the global integration value (GIV)

Table 4.1 shows that pedestrian numbers have the highest correlation with MSPA (0.311); this indicates a high correlation with measures related to proximity to adjacent facilities. However, the absolute value of all correlations was lower than 0.4; hence, a strong correlation was not shown. In addition, MSPS (+0.092), GIV (−0.131), and LIV3 (−0.047) did not show any correlation on their own although the research on the Nagoya Station underground mall complex reported that they were significantly related to pedestrian numbers.

4.4.2 *Regression Analysis Using the Stepwise Method*

On the basis of pedestrian numbers and the nine candidate explanatory variables, regression analyses using the stepwise method were conducted for up to three

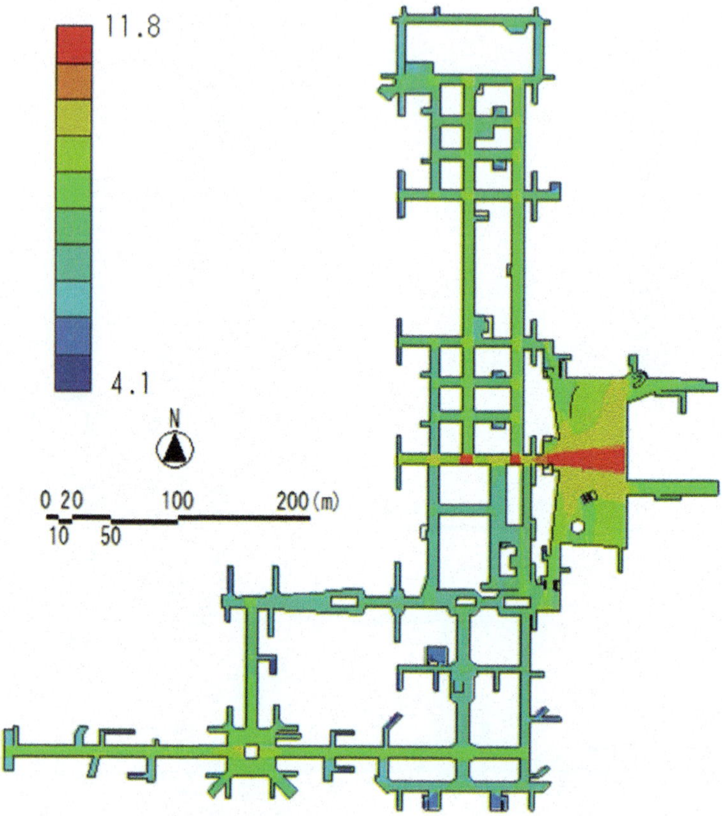

Fig. 4.10 Spatial distribution of the local integration value ($r = 3$) (LIV3)

variables. Of the resulting models, (1), (3), and (4) were found to have significant difference levels of 5% (Model (2) is included for reference.). However, the goodness-of-fit rankings that were based on the information criteria were Models (4), (3), and then (1). Table 4.2 shows the results of the regression analyses.

4.4.3 Discussion of the Models Using Regression Analysis

Model (1) followed a simple regression model using MSPA as the explanatory variable; the correlation coefficient was −0.311. When the simple regression models were examined, MSPA (−0.311) showed better goodness-of-fit than MSPS (+0.092)

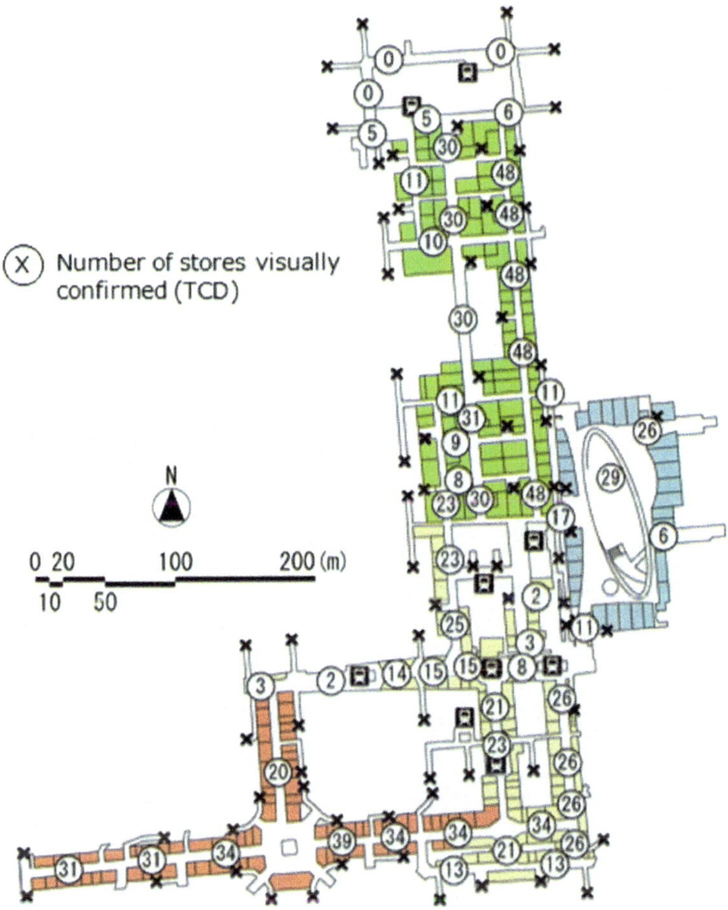

Fig. 4.11 Number of shops visually confirmed (TCD)

did; therefore, in the case of one variable, the physical distance to adjacent facilities was adopted as the factor.

In the case of Model (3), two variables, in the intensity order of factors, VSDS (the standardized partial regression coefficient, −0.375) and TCD (+0.366) were chosen. The correlation coefficient improved to +0.418 from −0.242 of VSDS alone. Both proximity to ticket gates and shops affected pedestrian numbers; however, visibility measures and not physical distance were adopted for both.

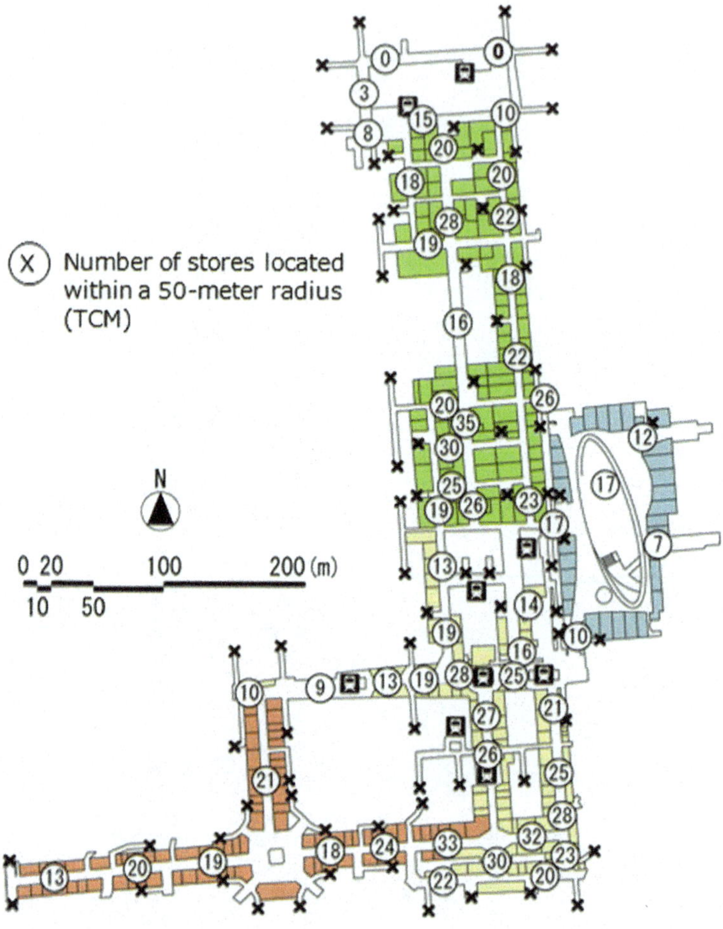

Fig. 4.12 Number of shops located within a 50-m radius (TCM)

In the case of Model (4), with three variables, in addition to VSDS (−0.415) and TCD (+0.385), VSDA (−0.288), which denotes the proximity to adjacent facilities, was adopted, and the correlation coefficient also improved to −0.507. At this time, all three adopted variables were visibility measures; this could be interpreted as higher proximity to ticket gates, facilities, and shops equating to higher pedestrian numbers.

Table 4.1 Correlation matrix between pedestrian numbers and candidate explanatory variables

		Connectivity	Ticket gate		Adjacent facility		Integration value		Tenant count	
	Pedestrian	CNT	VSDS	MSPS	VSDA	MSPA	GIV	LIV3	TCD	TCM
Pedestrian	1.000	–	–	–	–	–	–	–	–	–
CNT	−0.005	1.000	–	–	–	–	–	–	–	–
VSDS	−0.242	0.375	1.000	–	–	–	–	–	–	–
MSPS	0.092	0.169	0.515	1.000	–	–	–	–	–	–
VSDA	−0.235	−0.258	−0.116	−0.247	1.000	–	–	–	–	–
MSPA	−0.311	−0.019	0.121	−0.299	0.646	1.000	–	–	–	–
GIV	−0.131	0.332	−0.078	−0.311	0.365	0.302	1.000	–	–	–
LIV3	−0.047	0.810	0.494	0.395	−0.314	−0.096	0.180	1.000	–	–
TCD	0.230	0.493	0.364	0.409	0.013	−0.185	0.266	0.447	1.000	–
TCM	0.225	0.006	0.105	0.062	−0.018	0.028	0.290	−0.043	0.483	1.000

Table 4.2 Results of the multiple regression analyses

Model			Correlation coefficient		Standardized partial regression coefficient	AIC	Statistical test	
			Multi	Single				
(1)	Single	MSPA	–	−0.311	−0.311	629.743	[a]	
(2)	Single	VSDS	NA	−0.242	−0.242	632.016	[b]	
(3)	Multi	VSDS	0.418	−0.242	−0.375	626.755	[c]	[c]
		TCD		0.230	0.366			[c]
(4)	Multi	VSDS	0.507	−0.242	−0.415	623.013	[c]	[c]
		VSDA		−0.235	−0.288			[a]
		TCD		0.230	0.385			[c]

[a]<5%
[b]<10%
[c]<1%

4.5 Conclusion

Examining pedestrian numbers in the Sakae District underground mall complex, this research conducted multiple regression analyses by using visibility measures taken from SS theory and proximity measures of the underground shops. The results show that the model with the best goodness-of-fit, according to the information criterion, had three variables, with a correlation coefficient of 0.507, VSDS (−0.415), TCD (+0.385), and VSDA (−0.288). These results allow us to explain the pedestrian distribution of the Sakae District underground mall complex using proximity to adjacent facilities and shops, as well as proximity to the nearest station ticket gates, as factors. In addition, for all three variables, visibility measures were adopted; therefore, this research study also supports the validity of SS measures. This result contrasts with that of the Nagoya Station underground complex study, which reported that centrality (the integration value) in an underground complex configuration was a factor of crowding.

The analysis conducted in this research treated the whole underground area of Sakae District as one closed case; hence, an interesting future task would be to expand the scope of this analysis by considering through passages to the ground level and to establish a coupled case connecting underground sections to the ground level.

References

Afroza P, Arlen M, Beisi J (2007) Multilevel pedestrian movement: does visibility make any difference? In: Proceedings of the sixth international space syntax symposium, Istanbul, vol 040, pp 1–16

Araya R, Takeshita T, Ikezoe M (2005) Analysis of open space in urban area based on space syntax theory. J Architect Plann 589:153–160

Desyllas J, Duxbury E, Ward J, Smith A (2003) Pedestrian demand modelling of large cities: an applied example from London. UCL Centre for advanced spatial analysis working papers, 62

Hanazato T, Lee K, Takahashi T (1991) An introduction to space syntax theory: part 4. A case study in Yanaka and Hongo. Summaries of technical papers of annual meeting architectural Institute of Japan, architectural planning and design rural planning, pp 223–224

Hillier B, Burdett R, Peponis J, Penn A (1987) Creating life: or, does architecture determine anything? Architect Comport Architect Behav 3(3):233–250

Hillier B, Perm A, Hanson J, Grajewski T, Xu J (1993) Natural movement: or, configuration and attraction in urban pedestrian movement. Environ Plann B 20:29–66

Okamoto K, Kaneda T, Ota A, Meziani R (2013) Correlation analyses between underground spatial configuration and pedestrian flows by space syntax measures: a case study of an underground mall complex in Nagoya Station. In: 20th international seminar on urban form, pp 116–128

Ota A, Meziani R, Sugiura Y, Kaneda T (2008) A factor analysis on pedestrian numbers on paved streets in downtown using visibility indicators: a case study of the Sakae-South District, a Downtown Area of Nagoya. In: Lucas L, Mair G (eds) Sensory urbanism proceedings, pp 158–164

Ota A, Nakano Y, Kaneda T (2015) Comparative analysis on factors of the pedestrian numbers in a downtown area using space syntax indicators – case comparisons of Sakae-South, Nagoya CBD in Japan between 2005 and 2011. In: Proceedings of the tenth international space syntax symposium, London, vol 082, pp 1–16

Özer Ö, Kubat AS (2007) Walking initiatives: a quantitative movement analysis. In: Kubat AS, Ertekin Ö, Güney YI, Eyübolou E (eds) Proceedings of the Sixth International Space Syntax Symposium in Istanbul, ITU Faculty of Architecture, pp 017.1–017.16

Ueno J, Nakazawa A, Kishimoto K (2009) An analysis of pedestrian movement in multilevel complex by space syntax theory—in the case of Shibuya Station. In: Koch D, Marcus L, Steen J (eds) Proceedings of the seventh international space syntax symposium. Royal Institute of Technology, Stockholm, pp 118.1–118.12

Chapter 5
A Comparative Study of Factors of Land Price Index by Space Syntax Measures in Nagoya CBD Between 1935 and 1965: Case Study on Nagoya Downtown Area During Pre- and Post-war Period

Akira Ota and Toshiyuki Kaneda

Abstract The purpose of this research is to clarify the usefulness of the Space Syntax (SS) index to quantify street networks in a factor analysis for prosperity. (1) We focused on Nagoya CBD between 1935 and 1965, when the city planning projects, including the War Disaster Reconstruction Project, were executed and analyzed the land price index by multiple regression analysis using the distance from the nearest tram stop, commercial and office business accumulation, street width, and the SS index as factors in 1935 and 1965. (2) We compared the two estimated models, focusing on the changes in factor order. We concluded that the SS index has an impact on prosperity for city planning projects.

5.1 Introduction

5.1.1 Research Background and Objectives

When considering the construction of a city center from a long-term perspective, it is important to ascertain the reasons why certain sites are highly prosperous (this affluence hereafter referred to as "prosperity"), as well as to ascertain the enduring factors that affect the high degree of prosperity through changing eras in particular. Transitioning from one era to another causes changes in these factors, which may be affected by the

A. Ota (✉)
Keio University, Yokohama, Kanagawa, Japan

Tokyu Land Corporation, Tokyo, Japan
e-mail: akira_ota@keio.jp

T. Kaneda
Graduate School of Engineering, Nagoya Institute of Technology, Gokiso, Showa, Nagoya, Japan
e-mail: kaneda@nitech.ac.jp

T. Kaneda (ed.), *Downtown Dynamics*, Agent-Based Social Systems 16,
https://doi.org/10.1007/978-4-431-54901-7_5

context of that era as a matter of course, and by large-scale city planning projects that transform the appearance of street networks. To ascertain the change in these factors, an effective method involves conducting a comparative study of their influence in each era by analyzing the causes of prosperity in the downtown area.

Here, the indicators of prosperity are, in addition to the volume of pedestrian passage and congested traffic, land price, land rent, and property rent, which demonstrate the burden of land rent, and the public evaluation of land prices, which demonstrates the tax burden. This chapter focuses on land price, land rent, property rent, and public evaluations of land prices, for which information has been accumulated for a relatively long time. In this chapter, the land price, land rent, property rent, and public evaluations of land prices are referred to as a whole as the "land price index."

Looking at existing research reports that analyze the factors behind prosperity in Japan, a multiple regression model has been used for the land price index; however, these reports (e.g., Okubo 1983) deal with factors such as convenience of transport, including the distance from the nearest railway station, accumulation of purposes of nearby land use (referred to hereafter as "land use accumulation"), automobile transport accessibility, and street width that regulates maximum floor area ratios.

The UCL Group has developed a space syntax (hereafter, "SS") theory and has proposed a method for quantifying street network characteristics to analyze the factors behind prosperity, incorporating indicators such as street network centrality (hereafter, "centrality of street networks") and visible areas. SS theory is a spatial analysis theory that defines the phase relation, as advocated mainly by Bill Hiller at the University of London. In this study, a visual graph analysis is used based on a concept of viewing an entire area from a single point called "isovist" within SS theory. In the visual graph analysis, the target area is divided by detailed grids of any size with the summit as the center of each detailed grid, and a quantitative index is calculated such as integration value by converting the lines that connect each summit to a graph without crossing over between the covers. In this study, the target area was divided by a detailed grid of 4 m. In this chapter, these indexes are collectively referred to as the "SS index."

However, among the reports that analyze the factors behind land price indexes incorporating the SS index, none make a comparison between eras, which would be effective in ascertaining the enduring factors; neither do any of the reports consider the influence of large-scale street network revisions (hereafter, "street network revision") due to city planning projects. Graph theory is a method of quantifying the centrality of street networks; however, it is composed of nodes and links, so it is not possible to add spatial attributes to the data. Therefore, in this study, a discussion of street network centrality is made by means of SS theory, which is capable of considering spatial attributes using a detailed grid.

In this chapter, analysis is conducted targeting downtown Nagoya regarding the factors that formulated the land price index in pre-war 1935 and post-war 1965 ("two eras" combining both 1935 and 1965 hereafter) using, as candidate factor variables, the distance from the nearest Nagoya City Tram stop (hereafter, "tram stop") (hereafter, "distance from tram stop"), the nearby accumulation of commercial office use (hereafter, "commercial business accumulation"), and street width, as well as the SS index, and the Global Integration Value (hereafter, "GIV") indicating the centrality of street networks.

After compiling existing reports from Japan and other countries, which analyze the factors behind prosperity incorporating the SS index, a comparison was made of the spatial distribution on each index in downtown Nagoya and a multiple regression analysis was conducted. Regarding the two estimate eras, while cross-checking the background and city planning projects of each era, the validity of the factor analysis of the land price index incorporating the SS index was verified. An investigation was also performed of the implied effects on prosperity due to street network revisions based on city planning projects by comparing the order of factors according to the strength of the standardized partial regression factor (hereafter, "factor order"). The strength of the estimated results of the factor order is also investigated by cross-checking the results of the analysis of the model that gives consideration to the internal spatial correlation reported as an analytical model of the land price index.

5.1.2 *Prior Research on the Prosperity Factor Analysis Using the SS Index*

Research shows that prior to this study, in Japan, surveys on the prosperity factor analysis utilizing the SS index in central areas had been conducted based on reports from the Architectural Institute of Japan and the City Planning Institute of Japan. In other countries, surveys have been conducted based on reports from the Space Syntax Symposium and ISUF. In these surveys, the prosperity index was referred to as the land price index and the regression analysis that was positioned as an indicator of pedestrian passage volume was referred to as the prosperity factor analysis, whereby the candidate factor variables were organized as the conventional distance from the nearest station (ticket gate), land use accumulation, sidewalk (street) width, floor area ratio limits as well as the SS index with a focus on factor order (Table 5.1).

In the land price index, Min et al. (2009) conducted a single regression analysis of public land prices using the centrality of street networks on the SS index and reported a single correlation factor of 0.750. Later, Wang (2010) conducted a multiple regression analysis on public land prices using the centrality of street networks on the SS index and maximum floor area ratios as the factors, with the SS index as the main factor and the maximum floor area ratio as the second factor.

Furthermore, regarding pedestrian passage volume, internationally, Desyllas et al. (2003) reported on an analysis that included the visible area on the SS index as a factor. Among the reports in Japan, Araya et al. (2005) undertook an analysis using the centrality of street networks on the SS index and reported that the main factor was the SS index, after which Ueno et al. (2009) also reported that the main factor in a multi-layered analysis was the SS index. In analyses by Ota et al. (2008)

Table 5.1 Existing reports of the prosperity factor analysis in central areas using the space syntax (SS) index

Author	Survey year	Target distinct	Index for prosperity	Number of sample	Correlation coefficient	Factor order by standardized partial regression						
						Space syntax index			Other factor			
						Centrality of street network (integration value)	Visible area	Visual step depth	Conventional distance from the nearest station (ticket gate)	Land use accumulation	Sidewalk (street) width	Floor area ratio limits
Min et al.	–	Seoul City (approximately 100 ha)	Public land price[a]	418	0.750	No. 1 (0.750)	–	–	–	–	–	–
Wang	2009	Taichung City (approximately 10,000 ha)	Public land price[a]	150	0.692	No. 1[a,b]	–	–	–	–	×[c]	No. 2[a,b]
Desyllas et al.	1999–2001	London City (approximately 2500 ha)	Pedestrian passage volume[a]	237	0.906	–	O[a,d]	–	O[d]	O[d]	O[d]	–
Araya at al.	2001	Around Kashii station in Fukuoka prefecture (approximately 21 ha)	Pedestrian passage volume	26	–	No. 1[b]	–	–	No. 2[b]	×[c]	–	–
Ota et al.	2005	Around Sakae station in Aichi prefecture (approximately 21 ha)	Pedestrian passage volume	61	0.611	–	No. 4 (0.139)[e]	–	No. 1 (−0.341)[e]	No. 2 (0.260) (commercial) No. 3 (0.184) (office)[e]	–	–
Ueno et al.	2007	Around Shibuya station in Tokyo (approximately 8 ha)	Pedestrian passage volume[a]	44	0.909	×[c]	No. 2 (0.412)	No. 1 (−0.557)	No. 3 (−0.169)	–	–	–

Okamoto et al.	2012	Nagoya station underground mall in Aichi prefecture (approximately 7 ha)	Pedestrian passage volume	50	0.628	No. 2 (0.200)	×[c]	×[c]	No. 1 (−0.548)	–	–	–

[a]Used after logarithmic conversion
[b]Selected but standard partial regression is unknown, and the order depends on the relative importance of the *t*-statistic
[c]Rejected by statistical test
[d]Selected but standard partial regression is unknown
[e]*t*-test did not show 5% significance level

and Okamoto et al. (2013), the main factor was the distance from the nearest station or ticket gate, while the SS index, which was chosen for statistical testing, was the second factor or lower.

Based on the factor order in existing reports on the prosperity factor analysis, the distance from the nearest station (ticket gate) tends to occupy a high position on the list, but the SS index was also high in some reports. In addition, the fact that the SS index was chosen for statistical testing in some reports suggests the potential for a prosperity factor analysis that includes the SS index.

5.2 Comparative Eras and Example Target Districts

In modern downtown Nagoya, there are two main points in the large-scale revisions to street networks based on city planning projects. The first is the opening of Sakura Street in 1937 together with the relocation and reconstruction of Nagoya Station; Nagoya Station was relocated approximately 200 m north, after which Sakura Station was opened to the east, as reported by the Nagoya City Planning Bureau and the Nagoya Urban Institute incorporated foundation (1999). The second point is the large-scale expansion of the width of the streets based on land readjustment in the War Disaster Reconstruction Project conducted between 1946 and 1963; at this time, many wide streets in particular were opened, including the so-called 100-m roads of Hisaya-Odori, Wakamiya-Odori, Fushimi, and Sakura Streets, as reported by the editorial board of the *Planning Bureau War Disaster Reconstruction Journal* (1984). The War Disaster Reconstruction Project continued as a project for revival following the war even after 1964, but large-scale revisions to the street network in the downtown area were completed by 1963; this chapter does not therefore cover the War Disaster Reconstruction Project after 1964. The eras for comparison were set as 1935 and 1965, which were immediately prior to, and following, these revisions to the street network.

The example target district was set as a 3 km^2 area comprising the district east of Nagoya Station and the Sakae district, where these revisions to the street network could be seen, and which are the main areas of prosperity in Nagoya. As regards the borders on the north, south, east, and west, the west is bordered by railway lines including Nagoya Station, physically isolated by certain streets, while the borders on the north, south, and east comprise the relatively wide Sotobori Osu and Kuko Streets, where commercial office usage is decreasing. The analytical range of the SS index was calculated to be the same as that of the example target district. When setting the analytical range of the SS index, a special investigation was conducted so that there would be no major effect on the analytical results even if the analytical scope was changed.

5.3 Land Price Index and Candidate Factor Variables Used in the Analysis

5.3.1 Prior Research on the Prosperity Factor Analysis Using the SS Index

Maps with land price indexes for 1935 and 1965 include the cadaster (1:2400 scale) stated in the Nagoya land handbook of Ida (1934–1936) (referred to hereafter simply as “1935”) and the Route Price Setting Area Map (scale unknown) (hereafter, “route price map”) included in the 1965 inheritance tax property evaluation standards of the National Tax Agency, Direct Tax Department, and Asset Evaluation Planning Office (1965), based on which a polygon data street map (hereafter, “street map”) was produced for the example target district. However, because the scale of the route price map is unknown, the street map was produced using a method whereby corrections were made to sections that were different on the route price map based on the Nagoya City Commercial and Industrial Housing Map of the Housing Map Association (1965) (1:1500 scale) (hereafter, “residence map”). To obtain a certain number of sampling points with no deviation on the street maps produced for the two eras, 125 sampling points were established on a 200 m grid set up on the example target district (hereafter, “sample grid”). The positioning of the sampling points on each sample grid was, for 1935, the line at the center of the streets in front of the area with the highest land price index, and for 1965, the center of the streets with the highest land price index (Figs. 5.1 and 5.2). Based on the characteristics of each era, the land price index was applied to the land in 1935 and to the streets in 1965. Furthermore, when acquiring the land price index for 1935, the comparative 1965 land price index was the street value, and the street value was the estimated value of a standard interior lot, so corner lots and irregular plots were excluded from the 1935 land price index.

Comparing the street maps from the two eras, although there is no major difference in the positioning of tram stops, in the 1965 street map, the tram stop in front of Nagoya Station has moved to the north and the Sasashima-cho tram stop built near the 1935 Nagoya Station tram stop has changed due to the relocation of Nagoya Station. Moreover, due to the large-scale street width expansion based on the War Disaster Reconstruction Project, it was confirmed that in 1965 the streets were generally wider in comparison to 1935.

5.3.2 Extraction of Land Price Index and Candidate Factor Variables

Regarding the target land price index data, with a focus on the highest points of prosperity, the highest values on the sample grid for land rent prices in 1935 (Ida 1934–1936) and street values in 1965 (Asset Evaluation Planning, Direct Tax

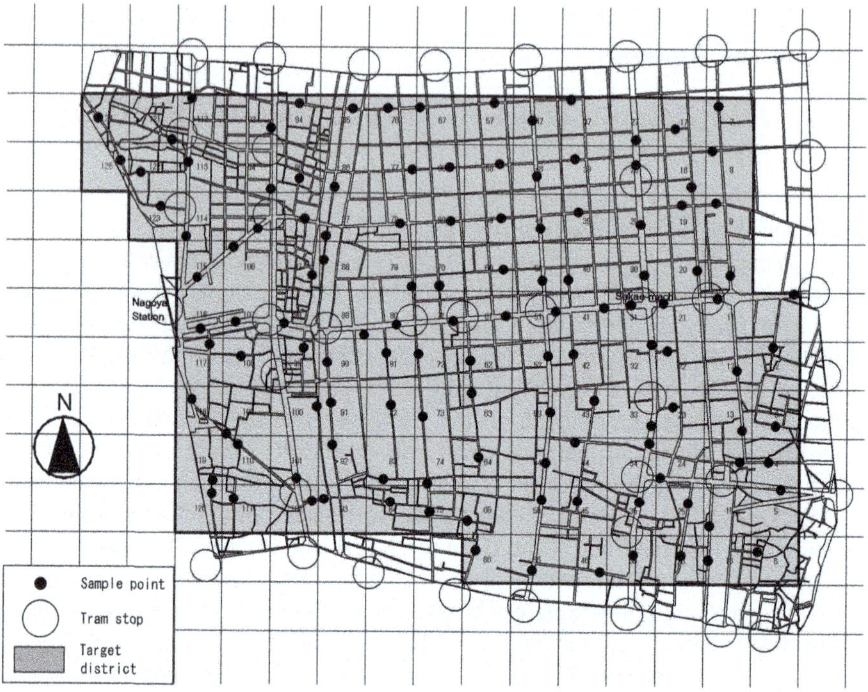

Fig. 5.1 1935 Nagoya downtown area street map and sampling points

Department Office, National Tax Agency 1965) were used. The land rent price refers to the price that forms the standard for taxation based on the studies conducted during the 1926 Act on Land Rent Price Investigations. A land survey was conducted, based on rental examples and the price of rice, and the grade was assessed in proportion to the land situation; the land rent price was determined through deliberations by a committee comprising members chosen by each town. The grade was converted into a monetary amount based on the grade table (hereafter, "land rent price grade table") defined by the former Ministry of Finance (now the Ministry of Finance and the Financial Services Agency) according to the quit-rent-related materials 2 (Tax Document Room, Tax Information Center, Tax University 2007) and the history of land rental price survey of the Nagoya Tax Supervision Bureau (1929). As 100 types of grade for each plot of land for land rent prices are stated in the handbook, the grade was converted to a land rent price based on the land rent price grade tables listed separately.

To ascertain the distances from tram stops, the position of the tram stop was plotted, and the distance of the street from the nearest tram stop to the sampling point was measured. The data from Takeuchi (1935) was used for 1935 and the data from the Housing Map Association (1965) was used for 1965.

Using the present situation of building use in Nagoya City from the City Planning Aichi Regional Committee (1938) and the City Planning Division, Nagoya City

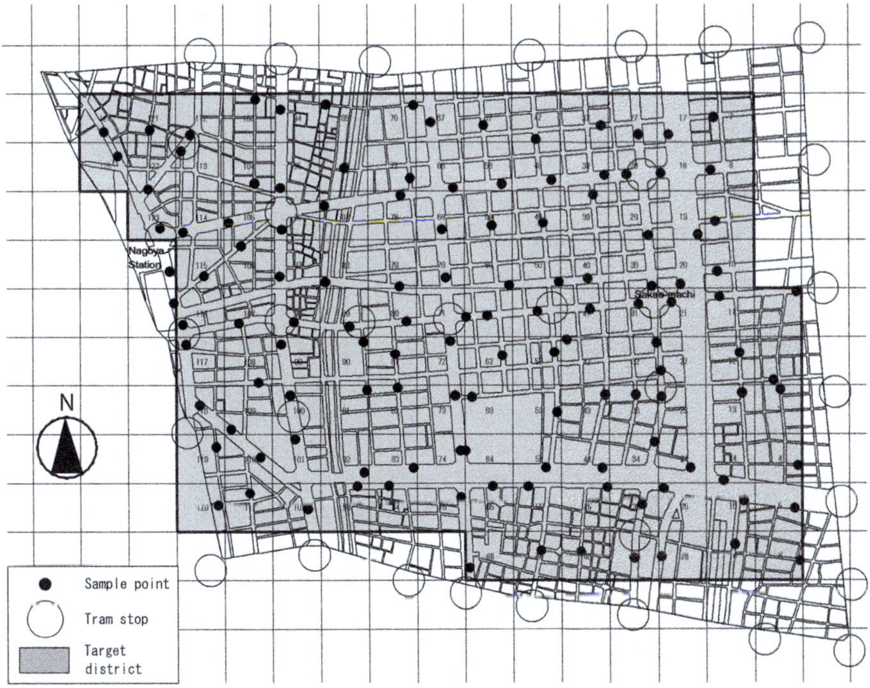

Fig. 5.2 1965 Nagoya downtown area street map and sampling points

Planning Bureau (1967), which were close to 1935 and 1965, commercial business accumulation was measured using the ratio of commercial and office use lot frontages on the boundary lines on both sides of the street line segments adjacent to the sampling points.

The street width at the sampling point was measured. Hisaya-Odori Street and Wakamiya-Odori Street include parks but are viewed as streets using the measurement of width and the SS index.

For the SS index, an index of the integration value ("IV" hereafter) was produced that shows the centrality of street networks using conceptual visible areas. The IV shows the strength of the spatial connection among street networks, with higher values having less depth, showing greater street network centrality. To find the IV, first, the target district is divided by grids, and the grid total is given as k. The depth of the graph comprising lines of view to the center of other grids from a central point on each grid as seen from the area as a whole is given as Depth, and the Mean Depth (hereafter, "MD") is found. Using MD and k, the relative asymmetry (hereafter, "RA") is found (Eq. 5.1). Next, as the RA is affected by the total of the grid, the Real Relative Asymmetry (hereafter, "RRA") is calculated (Eq. 5.3), which is the standardized value based on the correction factor D_k (Eq. 5.2) that determines k only. The reciprocal numbers that make the RR intuitively easier to understand are then given as the integration value (IV) (Eq. 5.4).

$$\mathrm{RA} = \frac{2(\mathrm{MD} - 1)}{k - 2} \quad (5.1)$$

$$D_k = \frac{2\left[k\left\{\log_2\left(\frac{k+2}{3}\right) - 1\right\} + 1\right]}{(k-1)(k-2)} \quad (5.2)$$

$$\mathrm{RRA} = \frac{\mathrm{RA}}{D_k} = \frac{(\mathrm{MD} - 1)(k - 1)}{\left[k\left\{\log_2\left(\frac{k+2}{3}\right) - 1\right\} + 1\right]} \quad (5.3)$$

$$\mathrm{IV} = \frac{1}{\mathrm{RRA}} \quad (5.4)$$

As this study covered the downtown area, using the GIV for the whole area, the GIV at the sampling points was measured as the SS index.

5.3.3 Spatial Distribution of Land Price Index and Candidate Factor Variables

The spatial distribution of the 1935 and 1965 land price indexes, commercial business accumulation of candidate factor variables, and GIV, respectively, are shown in Fig. 5.3. When producing a histogram for the land price index, the shape was close to that of lognormal distribution, so the values used were converted to logarithms.

Based on Fig. 5.3a, b, the land price index in 1935 shows a linear distribution of high values along Hirokoji, Otsu, and Honmachi Streets; however, in 1965, there is a non-linear distribution of high values near the Sakae-machi tram stop and the Nagoya Station tram stop.

Regarding commercial business accumulation, based on Fig. 5.3c, d, although roughly the same trends as those for the land price index can be seen, when focusing on the area around the Osu tram stop, for example, the land price index is relatively low in 1965 in comparison to 1935, yet there is no major disparity in commercial business accumulation between the two eras, so the two indexes show partial differences in the change patterns between the eras.

Based on Fig. 5.3e, f, in 1935, the highest GIVs are distributed on Hirokoji Street. This is followed by Otsu Street and Honmachi Street, showing the same trends as for the land price index. Meanwhile, in 1965, whereas high GIVs were distributed throughout due to revisions to the street network based on the War Disaster Reconstruction Project, high values in the land price index were concentrated around the Sakae-machi tram stop and the Nagoya Station tram stop in particular, showing that the land price index and GIV follow different patterns of change in distribution.

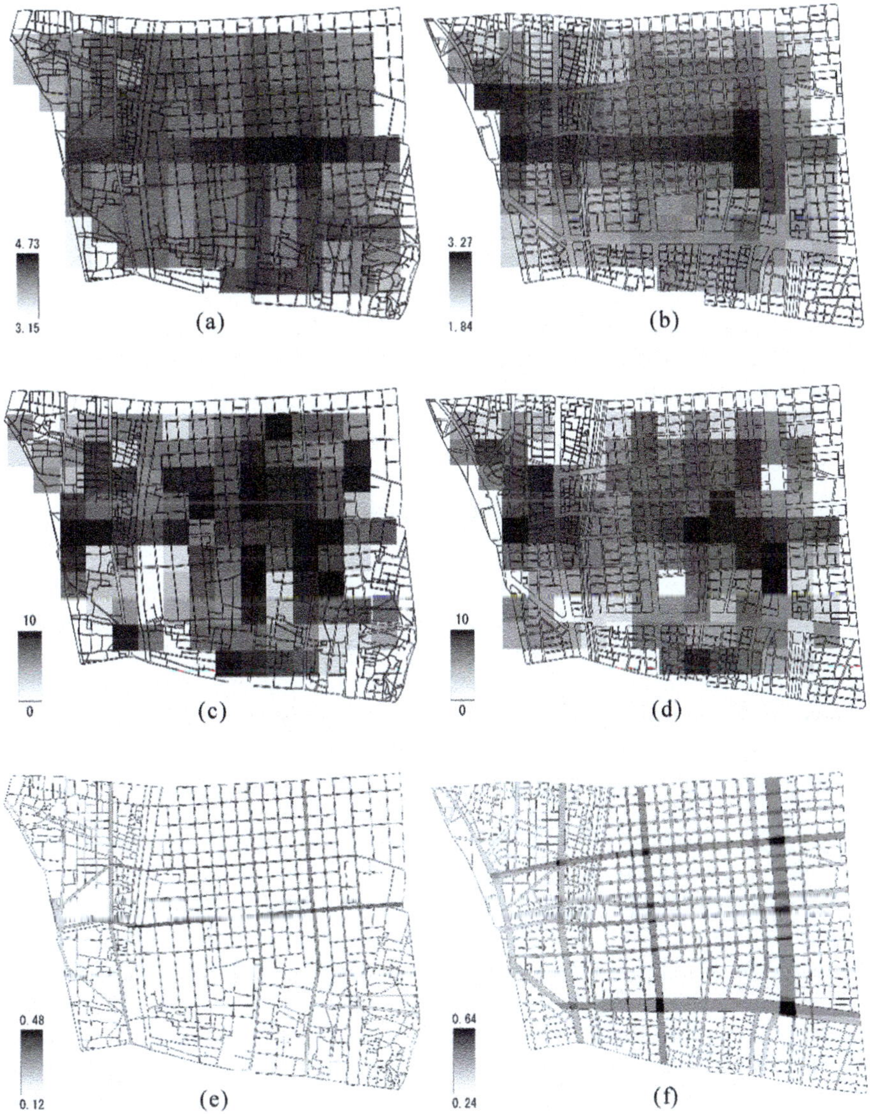

Fig. 5.3 Spatial distribution chart. (**a**) 1935 land price index; (**b**) 1965 land price index; (**c**) 1935 commercial business accumulation, (**d**) 1965 commercial business accumulation, (**e**) 1935 global integration value (GIV); (**f**) 1965 GIV

5.4 Investigation of Land Price Index Factors by Multiple Regression Analysis

5.4.1 Correlation Matrix Between Land Price Index Candidate Factor Variables

A correlation matrix for the 1935 and 1965 land price indexes and candidate factor variables was produced (Table 5.2). For the 1935, (Y2a)35_ln (land rent price) has a correlation factor of 0.699 and 0.657 with the (X2a)35_commercial business accumulation and (X4a)35_GIV, while for the 1965, (Y2b)65_ln (street value) has a correlation factor of 0.745 with (X2b)65_commercial business accumulation, showing a reasonably high correlation in each.

5.4.2 Study of the Model of Results from the Multiple Regression Analysis

Regarding the land price indexes in 1935 and 1965, the multiple regression equation was found by conducting a multiple regression analysis using a semi-logarithm model, and the suitability of that equation was investigated. With respect to the candidate factor variables, it was understood that there is no multicollinearity because the Variance Inflation Factor (hereafter, "VIF") between all variables was two or below, and so all were covered (Table 5.2).

In the *t*-test of the multiple regression model for the two eras, three variables were adopted after rejecting the 1935 (X1a)35_distance from tram stop and the 1965 (X3b)65_street width because the p values in the model for the four variables were 0.717 and 0.759, respectively. In addition, in the Akaike Information Criterion (hereafter, "AIC"), the lowest values were found for the three-variable model for both eras, so the three variables were adopted (Table 5.3). Tables 5.4 and 5.5 show the analytical results of the three-variable model, after which the three-variable model is discussed. Regarding the multiple regression model for 1935 prior to the street network revision, the multiple correlation factor was 0.808, and the factor variables were, in order of strength of the standardized partial regression factor, (X2a)35_commercial business accumulation (0.419), (X4a)35_GIV (0.308) and (X3a)35_street width (0.266). As regards the 1965 multiple regression model after the street network revision, the multiple correlation factor was 0.807, with (X2b)65_commercial business accumulation as the first factor (0.594), (X1b)65_distance from tram stop as the second factor (−0.295), and (X4b)65_GIV as the third (0205).

The obtained multiple regression model was compared with existing reports of prosperity factor analyses in central areas using the SS index in Tables 5.4 and 5.5.

In some existing reports, the SS index ranks high; however, this was not the case in this study for either of the two eras, being at No. 2 in 1935 and No. 3 in 1965.

Table 5.2 Correlation factor of land price index and candidate factor variables and VIF between candidate factor variables

VIF / Correlation	(Y1a)35_ land rent price	(Y2a)35_ln (land rent price)	(X1a)35_ distance from tram stop	(X2a)35_ commercial business accumulation	(X3a)35_ streets width	(X4a)35_ GIV	(Y1b)65_ roadside price	(Y2b)65_ln (roadside price)	(X1b)65_ distance from tram stop	(X2b)65_ commercial business accumulation	(X3b)65_ streets width	(X4b)65_ GIV
(Y1a)35_land rent price		–	–	–	–	–	–	–	–	–	–	–
(Y2a)35_ln (land rent price)	0.893		–	–	–	–	–	–	–	–	–	–
(X1a)35_distance from tram stop	-0.367	-0.326		1.065	1.369	1.051	–	–	–	–	–	–
(X2a)35_commercial business accumulation	0.607	0.699	-0.248		1.229	1.403	–	–	–	–	–	–
(X3a)35_streets width	0.640	0.592	-0.519	0.432		1.283	–	–	–	–	–	–
(X4a)35_GIV	0.574	0.657	-0.221	0.536	0.470		–	–	–	–	–	–
(Y1b)65_roadside price	0.619	0.484	-0.422	0.434	0.469	0.362		–	–	–	–	–
(Y2b)65_ln (roadside price)	0.593	0.550	-0.406	0.530	0.437	0.435	0.921		–	–	–	–
(X1b)65_distance from tram stop	-0.307	-0.236	0.681	-0.187	-0.369	-0.052	-0.487	-0.505		1.169	1.002	1.006
(X2b)65_commercial business accumulation	0.536	0.467	-0.336	0.553	0.402	0.347	0.656	0.745	-0.380		1.011	1.036
(X3b)65_streets width	-0.084	-0.006	0.096	0.033	-0.113	-0.083	-0.012	0.060	0.045	-0.105		1.566
(X4b)65_GIV	0.163	0.321	0.104	0.314	0.017	0.394	0.149	0.293	0.079	0.187	0.601	

Table 5.3 Multiple regression analysis: AIC test results

Era	Factor variables and its standardized partial regression	Correlation	AIC
1935	(X2a)35_commercial business accumulation: 0.699	0.699	−21.227
	(X2a)35_commercial business accumulation: 0.486 (X4a)35_GIV: 0.397	0.775	−50.220
	(X2a)35_commercial business accumulation: 0.419 (X4a)35_GIV: 0.308 (X3a)35_streets width: 0.266	0.808	−65.633
	(X2a)35_commercial business accumulation: 0.418, (X4a) 35_GIV: 0.309 (X3a)35_streets width: 0.254, (X1a)35_distance from tram stop : −0.023	0.808	−63.771
1965	(X2b)65_commercial business accumulation:0.745	0.745	−29.585
	(X2b)65_commercial business accumulation: 0.646, (X1b) 65_distance from tram stop : −0.260	0.783	−44.914
	(X2b)65_commercial business accumulation: 0.594, (X1b) 65_distance from tram stop: −0.295, (X4b)65_GIV: 0.205	0.807	−56.337
	(X2b)65_commercial business accumulation: 0.600, (X1b) 65_distance from tram stop: −0.293, (X4b)65_GIV: 0.191, (X3b)65_streets width: 0.022	0.808	−54.435

Table 5.4 1935 model multiple regression analysis results

	Regression coefficient	Standardized partial regression	*t* value	*p* value
Constant	3.325	–	46.252	0.000
(X2a)35_commercial business accumulation	0.004	0.419	6.410	0.000
(X4a)35_GIV	1.320	0.308	4.608	0.000
(X3a)35_streets width	0.012	0.266	4.253	0.000

Table 5.5 1965 model multiple regression analysis results

	Regression coefficient	Standardized partial regression	*t* value	*p* value
Constant	1.929	–	17.927	0.000
(X2b)65_commercial business accumulation	0.007	0.594	9.961	0.000
(X1b)65_distance from tram stop	−0.001	−0.295	−5.025	0.000
(X4b)65_GIV	0.779	0.205	3.704	0.000

However, as with existing reports, the SS index was chosen for statistical testing, and the validity of a prosperity factor analysis for downtown areas using the SS index was confirmed even in this study, which covered downtown Nagoya.

5.4.3 Consideration of Comparison Between Eras Focusing on the Factor Order

For the 1935 and 1965 models, a study of the factor order was conducted by making a comparison of the eras. While cross-checking with documents such as the *1935 37th Nagoya Statistical Record Book of Nagoya City Hall* (1937) and the *1965 56th Nagoya City Statistical Yearbook of Nagoya City General Affairs Bureau* (1966), an investigation was conducted of the implied effects on prosperity due to the context of the era and street network revisions based on city planning projects.

Commercial business accumulation was the first factor in both eras, and the strength of the standardized partial regression factor in 1965 was comparatively higher than that of 1935. In the statistical data, among the number of businesses by occupation in the center of the example target district, the occupation with the largest number of businesses in 1935 was manufacturing, whereas in 1965, it was the tertiary industry, which may be due to the increasing importance of commercial businesses.

Distance from a tram stop moved from being a rejected factor to being the second factor. Based on statistical data, the population of Nagoya approximately doubled between the two eras, while the number of passengers on the Nagoya City Tram increased approximately 26 times from some seven million in 1935 to 185 million in 1965, so there is a connection with the increased use of the Nagoya City Tram.

The GIV moved from being the second factor to being the third. As seen in the 1965 spatial distribution, this may have been influenced by the different GIV distribution situation in comparison to the land price index due to large-scale street network revisions based primarily on the War Disaster Reconstruction Project. However, even in 1965, the GIV was adopted at a significant level of 5% in the t-test; despite the decreasing effect of the centrality of street networks, it can be understood as retaining a certain level of influence.

Street width moved from being the third factor to being rejected. This can be seen from the fact that street width expanded rapidly due to the War Disaster Reconstruction Project, whereas that change did not greatly affect the land price index.

5.4.4 Study of the Model with Respect to Spatial Autocorrelation

The presence of spatial autocorrelation was confirmed because indexes such as those relating to land price are generally assumed to exert a mutual influence at connecting points due to spatial links. The adjoining connections were defined using a Delaunay triangulation comprising a triangular form that connects the points where the physical spatial distances are closest, and a spatial weight matrix was produced. Thereafter, based on Karato's (2007) "Estimation of a land price function and verification of spatial autocorrelation," a study of the model was conducted with respect to spatial autocorrelation. The calculation equations for various statistical figures are given by Karato (2007). A spatial econometric model was estimated by means of

Table 5.6 1935 spatial error model multiple regression analysis results

	Regression coefficient	Standardized partial regression	t value	p value
Constant	3.258	–	36.037	0.000
(X2a)35_commercial business accumulation	0.004	0.406	6.683	0.000
(X4a)35_GIV	1.551	0.362	4.672	0.000
(X3a)35_streets width	0.012	0.268	4.540	0.000

Table 5.7 1965 spatial lag model multiple regression analysis results

	Regression coefficient	Standardized partial regression	t value	p value
Constant	0.622	–	3.575	0.000
(X2b)65_commercial business accumulation	0.005	0.395	7.537	0.000
(X1b)65_distance from tram stop	−0.000	−0.229	−4.887	0.000
(X4b)65_GIV	0.744	0.196	4.475	0.000

Lagrange multiplier test statistics using the produced spatial weight matrix. A robust test was performed as the p value of LMλ, and LMρ in both 1935 and 1940 was significant (0.000). As a result, in 1935, LMλ∗ was 0.000 and LMρ∗ was 0.339, so LMλ∗ was significant, and a spatial error model could be estimated, while in 1965, LMλ∗ was 0.065 and LMρ∗ was 0.000, so LMρ∗ was significant, and a spatial lag model was estimated. The analytical results based on the estimated models are shown in Tables 5.6 and 5.7.

The AIC value was −94.065 in the 1935 spatial error model, and −102.64 in the 1965 spatial lag model, so both were lower than the model that does not give consideration to normal spatial autocorrelation; it was confirmed that the model that gives consideration to spatial autocorrelation is valid for both eras. In addition, the order of each variable factor in the model that gives consideration to spatial autocorrelation was the same as the model that does not.

5.5 Conclusion

In this chapter, regarding the formation of the land price index for 1935 and 1965 in downtown Nagoya, with the distance from a tram stop, commercial business accumulation, and street width taken as the candidate factor variables, a factor analysis was conducted that included the GIV showing the centrality of street networks on the space syntax (SS) index that quantifies street network characteristics, and a comparison was made of the factor order in an estimated model for the two eras.

1. Prior to the analysis, existing reports were compiled regarding the analysis of prosperity factors in the central area using the SS index. In those reports, a

regression analysis that positioned land price index and pedestrian passage volume as indicators of prosperity was conducted as a prosperity factor analysis, and a framework was presented in which the SS index was added to the candidate factor variables of the conventional distance from the nearest station (ticket gate), width of sidewalks (streets), and the maximum floor area ratio. The SS index occupies a high position in some existing reports. Furthermore, the fact that the SS index was chosen for statistical testing in some reports suggests that it can be added to the factors in a prosperity factor analysis.

2. After producing a polygon data street map based on maps with land price indexes for 1935 and 1965 in downtown Nagoya, the GIV data was extracted for distance from a tram stop, commercial business accumulation, street width, and the SS index as candidate factor variables, and the spatial distribution of land price index, commercial business accumulation, and GIV were presented. High GIVs were distributed on Hirokoji, Otsu, and Honmachi Streets in 1935, showing the same trend as that of the land price index. Meanwhile, in 1965, high GIVs were distributed throughout due to street network revisions based on the War Disaster Reconstruction Project; however, high values for the land price index were concentrated on the Sakae-machi tram stop area and the Nagoya Station tram stop area, which shows that the change in GIV distribution follows a different pattern to that of the land price index.
3. Multiple regression analysis was conducted for both 1935 and 1965 with distance from tram stop, commercial business accumulation, street width, and GIV as the candidate factor variables. In the *t*-test, three variables were adopted, with distance from tram stop in 1935 and street width in 1965 being removed. A three-variable model was used as three variables were adopted for the two eras in the AIC test. In the 1935 model prior to the street network revision, GIV followed commercial business accumulation as the second factor, with street width third, while in the post-street network revision 1965 model, GIV was the third factor, following commercial business accumulation and distance from tram stop. The SS index was chosen for statistical testing in both eras, in accordance with existing reports, and the validity was confirmed.
4. Regarding the estimated models for the two eras, with a focus on factor order, a comparison was conducted while cross-checking the era background and city planning projects based on statistical data. The placement of distance from tram stop moved from being a rejected factor to being the second factor; however, this is assumed to be connected to the increased use of the Nagoya City Tram. The placement of GIV moved from second to third, which is thought to be due to the influence of street network revisions based on city planning projects caused by changes in spatial distribution; however, even in 1965, the t test produced a significant level of 5%, and despite its diminishing effect, is deemed to have retained a certain level of influence. Moreover, in the model that gives consideration to spatial autocorrelation, AIC is lower than in the model that does not give such consideration, so the suitability of application was confirmed, but factor order was the same.

On this basis, regarding the validity of a prosperity factor analysis that includes the SS index as a factor in existing reports, even in the case that large-scale street network revisions are made due to city planning projects, in this study of downtown Nagoya, its effectiveness was confirmed to a certain degree. This insight suggests the possibility that, when considering long-term city center construction, it is valid to focus on the form of street networks in connection to the influence of city planning projects.

References

Araya R, Takeshita T, Ikezoe M (2005) Analysis of open space in urban area based on space syntax theory. J Architect Plann 589:153–160

Asset Evaluation Planning Office, Direct Tax Department, National Tax Agency (1965) Inheritance tax property evaluation standards. National tax agency

City Planning Aichi Regional Committee (1938) The present situation of building use in Nagoya city. City Planning Aichi Regional Committee

City Planning Division, Nagoya City Planning Bureau (1967) The present situation of building use in Nagoya city

Desyllas J, Duxbury E, Ward J, Smith A (2003) Pedestrian demand modelling of large cities: an applied example from London. Paper presented at UCL Centre for advanced spatial analysis working papers 62

Housing Map Association (1965) Nagoya City commercial and industrial housing guidebook map. Housing Map Association

Ida K (1934–1936) Nagoya land handbook Higashi-ku, Naka-ku and Nishi-ku. Empire of Japan municipal map publication

Karato K (2007) Estimation of a land price function and verification of spatial autocorrelation. The Land Institute of Japan

Min K, Moon J, Kim Y (2009) The effect of spatial configuration on land use and land value in Seoul. In: Proceedings of 7th international space syntax symposium, 63

Nagoya City General Affairs Bureau (1966) Nagoya City statistical yearbook 56, 1965. Nagoya City general affairs bureau statistical division

Nagoya City Office (1937) Nagoya City statistical report book, vol. 37. Nagoya City

Nagoya City Planning Bureau and Nagoya Urban Institute (1999) The history of Nagoya urban planning. Nagoya Urban Institute Incorporated Foundation

Nagoya Tax Supervision Bureau (1929) The history of land rental price survey business. Nagoya Tax Supervision Bureau

Okamoto K, Kaneda T, Ota A, Meziani R (2013) Correlation analyses between underground spatial configuration and pedestrian flows by space syntax measures: a case study of an underground mall complex in Nagoya Station. In: 20th International seminar on urban form, pp 116–128

Okubo M (1983) Land price and city planning. Gakugei Shuppan-sha

Ota A, Meziani R, Kaneda T (2008) A factor analysis on pedestrian numbers on paved streets in downtown using visibility indicators: a case study of the Sakae-South District, a downtown area of Nagoya. In: Sensory Urbanism Proceedings, pp 158–164

Planning Bureau War Disaster Reconstruction Journal Editorial Board (1984) The war disaster reconstruction journal. Nagoya City Planning Bureau

Takeuchi T (1935) Nagoya City map. Takeuchi Tokio

Tax Document Room, Tax Information Center, Tax University (2007) Land tax related materials 2. Finance Association

Ueno J, Nakazawa A, Kishimoto K (2009) An analysis of pedestrian movement in multilevel complex by space syntax theory—in the case of Shibuya Station. In: Koch D, Marcus L, Steen J (eds) Proceedings of the seventh international space syntax symposium. Royal Institute of Technology, Stockholm, pp 118.1–118.12

Wang W (2010) Using land value map as an alternative approach to analyze urban form: a preliminary study. In: 17th Conference international seminar on urban form, 2010

Chapter 6
Factor Analysis of Office Rent in the Area Around Kanda Station Using Space Syntax Theory: A Comparison with an Analysis of Shibuya Station

Akira Ota, Hiroshi Takahashi, and Toshiyuki Kaneda

Abstract When considering location—an important aspect—in real estate evaluation, the visible area indicates the location's visibility and the Integration Value (IV) indicates the location's ease of access. We believe that the quantification of real estate location evaluation can be advanced by analyzing rent factors using the visible area and the IV as Space Syntax (SS) indices. In this paper, we performed a multiple regression analysis on office rent, targeting the area within a 10-min walking distance of Kanda Station, Tokyo by adding the visible area and the IV to the other candidate factors. A model with the minimum value of the Akaike information criterion was selected using a stepwise variable increase/decrease method, and the fitness of the model was considered. The robustness of the SS index was also examined by comparing this paper to a previous study on Shibuya Station.

6.1 Introduction

6.1.1 Research Background and Objectives

Although location is important in real estate evaluation, its quantification is difficult, and the advantages and disadvantages of a location are normally discussed based on

---

A. Ota (✉)
Keio University, Yokohama, Kanagawa, Japan

Tokyu Land Corporation, Tokyo, Japan
e-mail: akira_ota@keio.jp

H. Takahashi
Keio University, Yokohama, Kanagawa, Japan
e-mail: htaka@kbs.keio.ac.jp

T. Kaneda
Graduate School of Engineering, Nagoya Institute of Technology, Gokiso, Showa, Nagoya, Japan
e-mail: kaneda@nitech.ac.jp

T. Kaneda (ed.), *Downtown Dynamics*, Agent-Based Social Systems 16,
https://doi.org/10.1007/978-4-431-54901-7_6

instinct. One possible method of location quantification is the space syntax (hereafter, "SS") theory advocated by Hillier and Hanson (1984) from the University of London. SS theory is used for the calculation of a city's spatial composition on a quantitative scale. In visibility graph analysis, an analytical method based on SS theory, the following two indices (hereafter, "SS index") are obtained: one for the visible area, which shows visibility, and one for the integration value (IV), which shows the centrality of street networks. When considering location in real estate evaluations, the visible area indicates the location's visibility and the IV indicates the location's ease of access (e.g., a hub that is easily accessible from many points). We believe that the quantification of real estate location evaluation can be advanced by analyzing rent factors using the SS index and by examining the potential for the SS index as a rent factor because location evaluation is generally assumed to be reflected in rent and land prices.

With respect to international research, Desyllas (1999) reported an analysis of office rent by adding the SS index to candidate factor variables in Berlin. In addition, Ota et al. (2018a) performed a factor analysis of office/residential/shop rent using the SS index. However, there are still only a small number of studies in the literature, and their robustness is disputable.

In this paper, we will conduct a multiple regression analysis of office rent within a 10-min walking distance from Kanda Station in Tokyo, after adding the visible area and the IV of the SS index to the floor, building age, contracted floor area, number of stories in the building, distance from the station, and construction dummy as well as by calculating the correlation of each. A model with the minimum value of the Akaike Information Criterion (hereafter, "AIC") is selected using the stepwise variable increase/decrease method, a consideration of the fitness of the model is made, and the order of factors is examined on the basis of the size of the absolute value of the t-value of the factor variables. In addition, because Kanda Station, the target of this paper, is characteristically an office area, while previous research on Shibuya Station in Japan targeted a complex area combining offices, residences, and shops, we will examine the robustness of the SS index by comparing this paper to a previous study.

6.1.2 Prior Research on Rent Factor Analysis Using the SS Index

On the one hand, internationally, Desyllas (1999) reported a factor analysis of office rent by adding the SS index to location attributes, as mentioned above. According to this report, in the period from 1991 to 1997 in West Berlin, in addition to the rental period, the IV, which is the SS index, was statistically significant ($t = 6.145$, $p < 0.001$), and the coefficient of determination was 0.663. Similarly, following the rental period, the IV was statistically significant ($t = 9.354$, $p < 0.001$) in all of Berlin and the coefficient of determination was 0.481, as reported by Desyllas (2000). As for the factor analysis of shop rent, Netzell (2010) performed an analysis using the SS index in central downtown Stockholm. In addition to the frequency of retail use, dummy price cut, and district dummy, the IV was statistically significant ($t = 2.3$, $p < 0.05$), and the coefficient of determination was reported as 0.65.

On the other hand, in Japan, Ota et al. (2018b) performed a correction analysis on the relationship between office rent and the SS index in the area around Kanda Station and reported a correlation factor for office rent and the visible area as 0.242 and a correlation factor for office rent and the IV as 0.229. Furthermore, in the same study, a factor analysis was performed on office/residence/shop rent in the area around Shibuya Station using the SS index. The coefficients of determination of the model were set to office (0.475), residence (0.305), and shop (0.367), and it was reported that the IV was statistically significant for all models (office: $t = 2.284$, $p < 0.05$; residence: $t = -2.155$, $p < 0.05$; shop: $t = 2.254$, $p < 0.05$). Other than the IV, the selected factors were building age, contracted floor area, distance from the station, and steel construction dummy for offices; building age and distance from the station for residences; and first-floor dummy, distance from the station, and building age for shops.

To some extent, international reports have shown the effectiveness of rent factor analysis using the SS index, but reports in Japan are limited to a single study. Moreover, the factor analysis performed in Japan targeted a complex area combining offices, residences, and shops. Therefore, in this paper, to examine the applicability in districts with other uses, we performed a factor analysis of office rent by targeting the area around Kanda Station, which is known to be an office area.

6.2 Rent Data as an Explained Variable

Examples of signed contracts have been extracted from the Real Estate Information Network Systems (hereafter, "REINS") and used as office rent data (hereafter, "rent data") in this study. In general, real estate information is shared among real estate agencies using REINS and provided to customers. The main customers targeted by real estate information in REINS are primarily individuals or small- to medium-sized enterprises, and signed contract examples comprise data on relatively small-sized office properties. One reason for this is that real estate information on large-sized office properties is often provided by major brokers that directly connect owners and tenants.

Small offices of less than 3 tsubo (about 10 m^2), which are assumed to be rental offices or shared workspaces, were excluded from the obtained signed contract examples. In addition, when there were multiple signed contract examples from the same building, the example with the lowest rent was used, whereas the others were excluded. The rent price used was rent per 3.3 m^2, calculated by adding maintenance fees and common charges to the rent and dividing by area (tsubo). Rent, maintenance fees, and common charges include consumption tax.

We obtained 127 samples for Kanda Station in a target year from June 1, 2015 to May 31, 2016. In comparison, 177 samples of office rent data were obtained in the previous study on Shibuya Station, which targeted 2 years from June 1, 2014 to May 31, 2016.

6.3 SS Index and Candidate Factor Variables

6.3.1 SS Index

In SS theory visibility graph analysis, a quantitative index is calculated by dividing the target area with a fine grid of an arbitrary size; the center of each fine grid is a vertex, and the index is calculated by converting it into a graph that has a straight line connected to each vertex without intersecting shields. In this study, the target area was divided into a fine grid of 4 m. The total number of grids that can be seen from the grid at a certain point represents the visible area. This is called connectivity in Turner's *Depthmap 4: A Researcher's Handbook* (2004).

In addition, an index of IV, which indicates the centrality of street networks, is created. The IV indicates the strength of a street network's spatial connection. If the value is high, the depth is shallow and the centrality of the street network is high. First, we considered the topological distance between certain points (grids) as depth. We took an average of the total value of the depth from a certain point to all other points, known as the Mean Depth (hereafter, "MD"), as per the SS methodology of Sayed et al. (2014). The topological distance here refers to the number of visible lines between grids (depth) and indicates the shortest number of steps required to move from one grid to another. Relative Asymmetry (hereafter, "RA") (6.1), which is the relative depth between each point (grid) seen from the entire area, is calculated using the MD. The higher the RA value is, the deeper the space is considered to be, and the more space must be passed through when moving. However, as RA is affected by k, which is the total number of grids, a calculation is made of the Real Relative Asymmetry (hereafter, "RRA"), which is a standardized value using the correction factor D_k (6.2) determined only by k (6.3). The reciprocal of this value is the IV (6.4).

The range of analysis can be defined by the IV. However, we decided to measure the global IV targeting all areas because this research covers a wide area (downtown area).

$$\mathrm{RA} = \frac{2(\mathrm{MD} - 1)}{k - 2} \tag{6.1}$$

$$D_k = \frac{2\left[k\left\{\log_2\left(\frac{k+2}{3}\right) - 1\right\} + 1\right]}{(k-1)(k-2)} \tag{6.2}$$

$$\mathrm{RRA} = \frac{\mathrm{RA}}{D_k} = \frac{(\mathrm{MD} - 1)(k - 1)}{\left[k\left\{\log_2\left(\frac{k+2}{3}\right) - 1\right\} + 1\right]} \tag{6.3}$$

$$\mathrm{IV} = \frac{1}{\mathrm{RRA}} \tag{6.4}$$

The visible area and the IV above were calculated using visibility graph analysis for streets within 2 km of Kanda Station. The signed contract examples are in the range of 800 m, but a buffer of 1200 m was added so that the target area covered

Fig. 6.1 Spatial distribution of the visible area within 2 km of Kanda Station

streets within 2 km, as the SS index tends to be high at the center of the target area. The spatial distribution maps of the obtained visible area and the IV are shown in Figs. 6.1 and 6.2. In addition, for comparison, the spatial distribution maps of the visible area and the IV from the previous study on Shibuya Station are shown in Figs. 6.3 and 6.4.

The spatial distribution maps of the visible area (Figs. 6.1 and 6.3) confirm that wide streets have good visibility and that the visible area is large because streets in the area around Kanda Station are in a grid pattern. In contrast to Shibuya Station, where the value is high only along National Route 246, around Kanda Station, there are streets with high values in various places.

The spatial distribution maps of the IV (Figs. 6.2 and 6.4) show the same trend as the visible area in which the value increases on wide streets. However, there are similar trends in Kanda Station and Shibuya Station, with differences between high value points and low value points being smaller in the IV compared to the visible area.

Fig. 6.2 Spatial distribution of the integration value within 2 km of Kanda Station

6.3.2 Other Candidate Factor Variables

With reference to the previous research on Shibuya Station, the candidate factor variables of the SS index other than the visible area and the IV were set as the floor, the contracted floor area, the number of stories in the building, the age of the building, the building construction dummy (steel construction/RC/SRC) as building attributes, and the distance from the station as a location attribute.

6.4 Consideration of Office Rent Factors

6.4.1 Correlation Between Office Rent and Candidate Factor Variables at Kanda Station

Calculations were made of the correlation factor and the Variance Inflation Factor (hereafter, "VIF") of office rent and candidate factor variables at Kanda Station (Table 6.1). Coloring was applied to values that exceed the absolute value (0.4) of

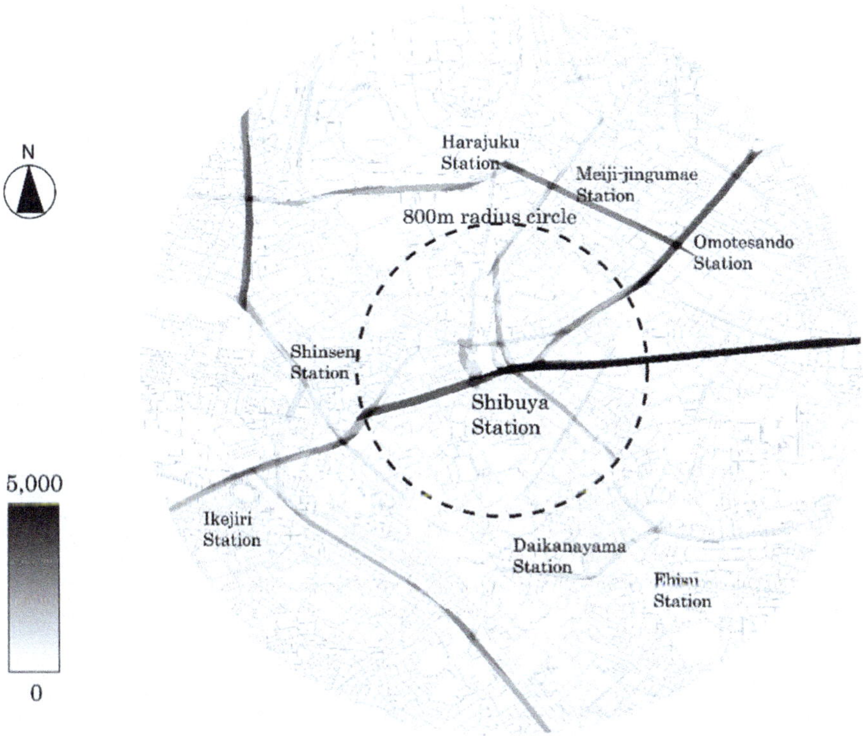

Fig. 6.3 Spatial distribution of the visible area within 2 km of Shibuya Station

the correlation factors, which was considered slightly high. VIF, which is considered a test index of multicollinearity, was calculated using the following expression (6.5), and coloring was applied to values over 2.

$$\mathrm{VIF} = \frac{1}{(1 - R^2)} \tag{6.5}$$

The only candidate factor variable that had a slightly high correlation with office property unit rent was building age (−0.485). Regarding the correlation between each candidate factor variable, the visible area and the IV (0.723) are highly correlated, and there is some correlation between the IV and the number of stories in the building (0.605), the floor number and the number of stories in the building (0.599), the RC dummy and the SRC dummy (0.594), the visible area and the number of stories in the building (0.579), the steel construction dummy and the SRC dummy (−0.468), and the visible area and the floor (0.445). As for VIF, the combination of visible area and IV exceeds 2, suggesting multicollinearity.

Fig. 6.4 Spatial distribution of the integration value within 2 km of Shibuya Station

Furthermore, the correlation between office property unit rent and candidate factor variables was compared to the previous research on Shibuya Station (Table 6.2).

Building age was ranked first at Kanda Station and second at Shibuya Station, and the contracted floor area was ranked third at Kanda Station and first at Shibuya Station, confirming that they were ranked higher at both stations. The IV, which is the SS index, was ranked fifth at Kanda Station and fourth at Shibuya Station; it is assumed that there is a certain degree of connection. The visible area was ranked fourth at Kanda Station, which is relatively high, but it was ranked the lowest at Shibuya Station; this is likely because of the extremely high value along National Route 246 at Shibuya Station. Moreover, the order of the number of stories in the building differed greatly between the two stations, ranking second at Kanda Station and eighth at Shibuya Station. This suggests that the number of stories in a building has an influence on office rent. Meanwhile, the distance from the station was ranked third at Shibuya Station and seventh at Kanda Station. There are many tenants who prefer the adequate accessibility at Shibuya Station, which is thought to lead to an increase in office rent.

Table 6.1 Correlation factor and VIF of office rent and candidate factor variables at Kanda Station

Correlation\VIF	(Y) Office rent (Yen/tsubo)	(X1) Distance from the station	(X2) Visible area	(X3) Integration value	(X4) Floor number	(X5) Contracted floor area	(X6) Number of stories	(X7) Building age	(X8) Building construction dummy (steel)	(X9) Building construction dummy (RC)	(X10) Building construction dummy (SRC)
(Y) Office rent (Yen/tsubo)	\	1.016	1.062	1.055	1.047	1.114	1.180	1.308	1.010	1.013	1.004
(X1) Distance from the station	−0.125	\	1.004	1.002	1.000	1.019	1.006	1.018	1.004	1.005	1.010
(X2) Visible area	0.242	0.060	\	2.094	1.247	1.000	1.505	1.012	1.005	1.070	1.099
(X3) Integration value	0.229	−0.049	0.723	\	1.124	1.004	1.577	1.002	1.005	1.039	1.097
(X4) Floor number	0.212	−0.019	0.445	0.333	\	1.011	1.560	1.048	1.016	1.040	1.014
(X5) Contracted floor area	0.320	0.137	0.020	0.064	0.103	\	1.028	1.016	1.000	1.001	1.003
(X6) Number of stories	0.390	0.075	0.579	0.605	0.599	0.166	\	1.064	1.005	1.047	1.113
(X7) Building age	−0.485	−0.132	−0.110	−0.039	−0.214	−0.126	−0.246	\	1.104	1.018	1.025
(X8) Building construction dummy (steel)	0.099	−0.065	−0.070	−0.071	0.127	−0.010	−0.072	−0.307	\	1.164	1.280
(X9) Building construction dummy (RC)	−0.113	−0.068	−0.255	−0.193	−0.197	−0.033	−0.211	0.131	−0.376	\	1.545
(X10) Building construction dummy (SRC)	0.063	0.098	0.300	0.297	0.118	0.058	0.318	0.155	−0.468	−0.594	\

Table 6.2 Correlation coefficient between office property unit rent and candidate factors (comparison between Kanda Station and Shibuya Station)

	Kanda Station		Shibuya Station	
Rank	Candidate factors	Correlation	Candidate factors	Correlation
1	(X7) Building age	−0.485	(X5) Contracted floor area	0.459
2	(X6) Number of stories	0.390	(X7) Building age	−0.424
3	(X5) Contracted floor area	0.320	(X1) Distance from the station	−0.323
4	(X2) Visible area	0.242	(X3) Integration value	0.235
5	(X3) Integration value	0.229	(X4) Floor number	0.191
6	(X4) Floor number	0.212	(X8) Building construction dummy (steel)	0.189
7	(X1) Distance from the station	−0.125	(X9) Building construction dummy (RC)	−0.184
8	(X9) Building construction dummy (RC)	−0.113	(X6) Number of stories	0.168
9	(X8) Building construction dummy (steel)	0.099	(X2) Visible area	0.101
10	(X10) Building construction dummy (SRC)	0.063	–	–

Table 6.3 Results of multiple regression analysis on office property unit rent at Kanda Station

Coefficient of determination: 0.417		AIC: 2364.38		
	Partial regression coefficient	Standard partial regression coefficient	*t* value	*p* value
Constant	14,360	–	10.999	0.000
(X7) Building age	−113	−0.420	−5.828	0.000
(X6) Number of stories	420	0.262	3.643	0.000
(X5) Contracted floor area	42	0.256	3.616	0.000
(X1) Distance from the station	−458	−0.235	−3.351	0.001

6.4.2 *Factor Analysis of Office Rent at Kanda Station*

Multiple regression analysis was performed on office property unit rent using the candidate factor variables. The VIF of the combination of visible area and IV among the candidate factor variables exceeded 2, which suggests multicollinearity. Therefore, first, we excluded the IV from the candidate factor variables and performed an analysis while retaining the visible area; second, we performed an analysis replacing the visible area with the IV. A model with the minimum value of the AIC was selected using a stepwise variable increase/decrease method.

Table 6.4 Results of multiple regression analysis on office property unit rent at Shibuya Station

Coefficient of determination: 0.475		AIC: 3353.37		
	Partial regression coefficient	Standard partial regression coefficient	*t* value	*p* value
Constant	18,430	–	0.000	1.000
(X7) Building age	−144	−0.380	−6.554	0.000
(X5) Contracted floor area	81	0.359	6.277	0.000
(X1) Distance from the station	−494	−0.293	−5.015	0.000
(X3) Integration value	23,414	0.134	2.284	0.024
(X8) Building construction dummy (steel)	1791	0.120	2.108	0.036

The same results were obtained from the analysis using the IV and that using the visible area (Table 6.3). The results from the previous research on Shibuya Station are shown in Table 6.4 for the purpose of comparison.

The coefficient of determination of the model for Kanda Station was 0.417; in order of the absolute value of the *t*-value, the factors were the building age, number of stories in the building, contracted floor area, and distance from the station. The results show that office rent tends to increase when the building is younger, the number of stories in the building is larger, the contracted floor area is larger, and the distance from the station is smaller. In this analysis, both the visible area and IV, which are the SS indices, were rejected as a result.

The coefficient of determination of the model is lower than that of the previous study on Shibuya Station, which was 0.475. While the IV and steel construction dummy, which were selected at Shibuya Station, were rejected, the number of stories in the building was selected at Kanda Station.

In this analysis on Kanda Station, while the SS index was rejected, to study the reasons for the selection of the number of stories in the building, we performed a multiple regression analysis. The IV and visible area were individually added to the building age, number of stories in the building, contracted floor area, and distance from the station, which are the factor variables of the model (Tables 6.5 and 6.6).

Based on Tables 6.5 and 6.6, the coefficient of determination slightly improved in the model with the SS index added to the factors compared to Table 6.3. However, the AIC worsened slightly, and the IV and visible area, which are the SS indices, showed no significant difference at 5% in the *t*-test. In addition, the number of stories in the building, which showed a significant difference at 1% in the *t*-test, did not show a significant difference at 1% in the model with the SS index added to the factors, and the factor order dropped from second to fourth. Although the VIF values of the number of stories in the building and the SS index are lower than 2 for both the IV and the visible area (1.577 and 1.505, respectively), a correlation factor of 0.605 was found between the number of stories and the IV, whereas a correlation factor of 0.579 was found between the number of stories in the building and the visible area, showing a somewhat high correlation. Therefore, the influence seems to be reciprocal. Consequently, the number of stories in the building was excluded, and a model

Table 6.5 Results of multiple regression analysis on office property unit rent at Kanda Station (integration values were added to the factors)

Coefficient of determination: 0.419		AIC: 2366.14		
	Partial regression coefficient	Standard partial regression coefficient	*t* value	*p* value
Constant	13,425	–	5.689	0.000
(X7) Building age	−114	−0.424	−5.823	0.000
(X5) Contracted floor area	42	0.257	3.614	0.000
(X1) Distance from the station	−452	−0.232	−3.275	0.001
(X6) Number of stories	377	0.235	2.565	0.012
(X3) Integration value	4229	0.042	0.476	0.635

Table 6.6 Results of multiple regression analysis on office property unit rent at Kanda Station (visible area was added to the factors)

Coefficient of determination: 0.422		AIC: 2365.44		
	Partial regression coefficient	Standard partial regression coefficient	*t* value	*p* value
Constant	14,694	–	10.862	0.000
(X7) Building age	−113	−0.422	−5.857	0.000
(X5) Contracted floor area	43	0.263	3.688	0.000
(X1) Distance from the station	−463	−0.238	−3.382	0.001
(X6) Number of stories	342	0.214	2.424	0.017
(X2) Visible area	0.345	0.081	0.948	0.345

with the minimum AIC values was selected for a multiple regression analysis using a stepwise variable increase/decrease method. As the IV and visible area are also suspected to exhibit multicollinearity, an analysis was performed by switching between each (Tables 6.7 and 6.8).

Based on Tables 6.7 and 6.8, compared to Table 6.3, the coefficient of determination and the AIC worsened slightly in the model that excludes the number of stories from the factor candidate. However, the IV and the visible area showed a significant difference of 5% in the *t*-test, suggesting that the two can be factors in office rent.

In contrast to the results of the previous study on Shibuya Station, the correlation between the number of stories in the building and the SS index was slightly high. In addition, because of the reciprocal influence in this analysis on Kanda Station, the SS index was rejected, and the number of stories in the building became the model for selection. In terms of the characteristics of Kanda Station, the maximum regulated floor area ratio is high along areas with good visibility, such as main streets and

Table 6.7 Results of multiple regression analysis of office property unit rent at Kanda Station (excluding the number of stories in the building and visible area from candidate factors)

Coefficient of determination: 0.387		AIC: 2370.87		
	Partial regression coefficient	Standard partial regression coefficient	*t* value	*p* value
Constant	12,192	–	5.160	0.000
(X7) Building age	−127	−0.472	−6.550	0.000
(X5) Contracted floor area	46	0.279	3.865	0.000
(X1) Distance from the station	−422	−0.217	−3.002	0.003
(X3) Integration value	18,273	0.182	2.556	0.012

Table 6.8 Results of multiple regression analysis of office property unit rent at Kanda Station (excluding the number of stories in the building and integration value from candidate factors)

Coefficient of determination: 0.394		AIC: 2369.46		
	Partial regression coefficient	Standard partial regression coefficient	*t* value	*p* value
Constant	17,016	–	17.468	0.000
(X7) Building age	−123	−0.458	−6.367	0.000
(X5) Contracted floor area	48	0.291	4.065	0.000
(X1) Distance from the station	−463	−0.238	−3.312	0.001
(X2) Visible area	0.854	0.200	2.821	0.006

intersections, as well as locations where the centrality of the street network is high. It is possible that there are the same trends for the SS index and the number of stories in the building. Thus, depending on the characteristics of an area, the number of stories might replace real estate location evaluation factors such as the visibility and centrality of the street network. Generally, the number of stories in a building indicates the building size, and offices in large buildings seem likely to have higher rent owing to extensive common areas and facilities. Therefore, it is thought that the number of stories in the buildings around Kanda Station was selected because of its superiority over the SS index, as it replaced the attributes of visibility and centrality of the street network as well as that of building size.

6.5 Conclusion

In this study, we performed a multiple regression analysis on office rent, targeting the area within a 10-min walking distance of Kanda Station in Tokyo by adding the visible area and IV, which are the SS indices, to the floor, building age, contracted floor area, number of stories, distance from the station, and construction dummy. A model with the minimum value of the AIC was selected using a stepwise variable increase/decrease method, the fitness of the model was considered, and the order of factors was examined from the size of the absolute value of the *t*-value of factor variables. In addition, we examined the robustness of the SS index by comparing this paper to a previous study on Shibuya Station.

The coefficient of determination of the model for Kanda Station was 0.417. The factors were, in order of the absolute *t*-value, the building age, the number of stories, the contracted floor area, and the distance from the station. The results showed that office rent tends to increase when the building is younger, when the building has more stories, when the contracted floor area is larger, and when the building is closer to the station. Moreover, both the visible area and the IV were rejected in this analysis. Further, the coefficient of determination of the model herein was lower than that of Shibuya Station, which was 0.475. Although the IV and steel construction dummy, which were selected at Shibuya Station, were rejected, the number of stories in the building was selected at Kanda Station.

Subsequently, we performed a multiple regression analysis by individually adding the IV and visible area to the building age, number of stories in the building, contracted floor area, and distance from the station. Although the coefficient of determination improved slightly, the AIC worsened somewhat, and the IV and visible area showed no significant difference of 5% in the *t*-test. Additionally, the number of stories in the building, which showed a significant difference of 1% in the *t*-test, did not show a significant difference of 1% in the model with the SS index added to the factors, and the factor order dropped from second to fourth. As the correlation between the number of stories in the building and the SS index is slightly high, it seems that there was a reciprocal influence.

Finally, the number of stories in the building was excluded, and a model with the minimum value of the AIC was selected using a multiple regression analysis via a stepwise variable increase/decrease method. Although the coefficient of determination and the AIC worsened somewhat, both the IV and the visible area showed a significant difference of 5% in the *t*-test, suggesting that IV and visible area can be factors in office rent.

Unlike the results of the previous study on Shibuya Station, in locations where the visibility and centrality of street networks were high, which is a characteristic of Kanda Station, the maximum regulated floor area ratio was high; it is thought that the SS index was rejected and the number of stories in the building was selected because the trend was the same for the number of stories in the building and the SS index. This confirmed the possibility that depending on the characteristics of the area, the number of stories in a building replaces real estate location evaluations such as the visibility and centrality of the street network.

To clarify the potential of the SS index for real estate location evaluations and its relationship with other factors, more analyses of a similar kind are required for other areas of Japan, particularly of areas with different characteristics.

References

Desyllas J (1999) Why pay to be there? In: Proceedings of 1st international space syntax symposium, 01, pp 1–12

Desyllas J (2000) The relationship between urban street configuration and office rent patterns in Berlin. University College of London, London

Hillier B, Hanson J (1984) The social logic of space. Cambridge University Press, Cambridge

Netzell O (2010) The effect of accessibility on retail rents: testing integration value as a measure of geographic location. J Prop Res 30(1):1–23

Ota A, Takahashi H, Kaneda T (2018a) Factor analysis of rent using space syntax measures: comparative analysis by building use around Shibuya Station. Jpn J Real Estate Sci 31 (4):109–118

Ota A, Takahashi H, Kaneda T (2018b) Analysis of the relation between office rent and evaluation of location using space syntax theory: a case study of the Kanda Station. In: Proceedings of 8th the Japanese Society for Artificial Intelligence Special Interest Group on Business Informatics

Sayed K, Turner A, Hillier B, Iida S, Penn A (2014) Space syntax methodology, 4th edn. Bartlett School of Architecture, UCL, London

Turner A (2004) Depthmap 4: a researcher's handbook. Bartlett School of Graduate Studies. University College London, London

Part III
Shopper Agents and Downtown Dynamics

Chapter 7
ASSA: Agent-Based Simulation Model for Shop-Around Agent Model

Takumi Yoshida

Abstract In this chapter, I summarize previous research on models for shop-around behavior and demonstrate the significance of developing the agent-based model for shop-around behavior. Before going into detail about the model, a definition of shop-around behavior is given that focuses on its characteristics, planned actions, and improvised actions. Shop-around behavior is broken down into two axes: time sequence and processing target. I then describe the development of the agent-based pedestrian model that has a bounded rational planning function, an autonomous correction function, and a strengthening learning function that updates shop preferences based on walk-around results. The basis of the definition for shop-around behavior is "multi-purpose multi-stop." I also present evaluation framework for this model consisting of five items in total, two for general evaluation and three for individual function evaluation.

7.1 Significance of Developing This Model

7.1.1 Definitions of Shop-Around Behavior

Before developing a shop-around behavior model, I will first define the characteristics of such behavior. The basis of the definition for shop-around behavior is "multi-purpose multi-stop," which includes the characteristics of O'Kelly (1981).

In Japan, it is defined by Yuzawa et al. (1993) as a course of behavior whereby, "for example, when going to a commercial district by car a certain district is first selected for the purpose of parking, then another district is visited for shopping... followed by shopping around, and returning to the first district for the purpose of returning home." Furthermore, Saito (2000) defines it as "consumer behavior to work around commercial facilities while changing their destination and purpose." The common points in both are the focus on commercial facilities, such as shopping, food and drink, or entertainment facilities and the repeated purposeful movement

T. Yoshida (✉)
Bureau of Housing and City Planning, Nagoya City Office, Nagoya, Japan
e-mail: t.yoshida.dr@city.nagoya.lg.jp

T. Kaneda (ed.), *Downtown Dynamics*, Agent-Based Social Systems 16,
https://doi.org/10.1007/978-4-431-54901-7_7

between facilities. Kaneda et al. (2001) consider shop-around behavior to have three steps: "Potential visitors to the downtown area determine the commercial district to visit according to their purpose in STEP 1. Visitors to the downtown area who have arrived at the commercial district choose the type of facility for the purpose, such as shopping, in STEP 2 and then choose which zone of the facilities to visit in STEP 3. They walk around from one facility to another by repeating STEP 2 and STEP 3, which results in shop-around behavior." While the first two definitions consider shop-around behavior to occur only in commercial districts, Kaneda et al. have a more comprehensive definition, as they view shop-around behavior as the entire series of steps encompassing decision-making and behavior from the planning to returning home, including decision-making at home. In this report, I define shop-around behavior in accordance with this definition by Kaneda et al. (2001).

Shop-around behavior is a course of behavior in which customers who have decided to visit commercial districts according to their errands choose a district and facilities and develop action plans, then visit the commercial district and walk around the commercial facilities with the purpose of satisfying various needs (such as exploration and rest) to finish their errand(s), before returning home. Furthermore, it should be noted that, in actual behavior, customers follow the developed plan, but often change it flexibly and repeatedly according to the situation. In the development of a model, I assume that it is necessary to focus on planned action and improvised action in shop-around behavior.

7.1.2 Shop-Around Behavior Models in Existing Research

The pedestrian shop-around behavior models in existing research are mainly aggregated models developed on the premise that shop-around behavior has the Markov property. International examples include the model by Borgers and Timmermans (1986), while those from Japan include the model by Sakamoto (1984) that used an absorbing Markov chain model, a model by Saito and Ishibashi (1992) based on the transition probability matrix between nodes and a logit model, and a model by Kaneda et al. (2001) that used transition probability among facility types and the Huff model.

The Markov property indicates that all information related to behavioral decision-making at the next point of time t_1 that occurs at a certain time t_0 is aggregated at the state in t_0. When developing a model for shop-around behavior using this concept, behavior study data-based fitting is relatively easy because behavior selection is expressed by state transition probability with the current situation as a parameter. Furthermore, it is possible to use special interaction models that deal with single errands/single visits, such as the Huff Model and Logit Model; in fact, models using Markov chains have been developed using such models.

However, these aggregated models have a drawback in that they cannot express the continuity of visits. Since shop-around behavior characteristically involves walking around facilities, the sequence itself makes sense because of individual pedestrians making decisions. Yet the aggregated models can express the sequence only as a result of multiplying state transition probability because the number of people is calculated according to a probability based on the ratio of study data. As a result, there is no point in focusing on individual sequences.

Therefore, in recent years, there has been a trend in developing non-aggregated models in shop-around behavior model research. This includes one by Mibuta et al. (2000a, b) that used an agent-based approach, and one by Kurose et al. (2001) that used heuristic selection rules. Both are application examples of complex simulation model technology that has recently been developed in the computer science field, a field wherein model construction can be expected to expand widely as technology improves in the future.

7.1.3 The Significance of Developing an Agent-Based Model for Shop-Around Behavior

I will now describe the significance of developing a shop-around behavior model using an agent-based approach.

A model developed through an agent-based approach (hereinafter, "agent-based model") uses a simulation method that makes calculations based on individuals called agents. Therefore, the characteristics are such that it is possible to provide selection rules for the knowledge and actions of each agent, and for each agent to autonomously make decisions according to their environment. This enables selection based on the purpose and history of each individual agent, and the limits of conventional models can be overcome.

In addition, unlike aggregated models, since the agent-based model takes a so-called bottom-up approach, which expresses macro-level social phenomena as an accumulation of the behavioral results of agents, the meaning of micro-level behavioral results is not necessarily the same as that of aggregated results. Conversely, it may be effective to focus on micro-level behavioral sequences. In other words, the agent-based model performs micro-simulation and macro-simulation at the same time, and the micro–macro link in social phenomena can be expressed and explicated.

Moreover, the need for agent-based modeling for shop-around behavior is encouraged by the diversification of consumer behavior because shop-around behavior reflects the diversified facility preference patterns of individual customers, and the agent-based model, in which each agent has a unique behavior rule, is suitable for expressing diversified preferences. Furthermore, agents can simulate changes in visitor behavior patterns caused by real-time changes in the setting of the commercial district because they learn and adapt to the environment according to the progress of the simulation. In addition, by describing not only visitors but also facilities as agents, it is possible to express the fact that facilities autonomously change through their interaction with visitors. It may, therefore, be possible to simulate a phenomenon in which a city develops and revises itself to become attractive, which has been impossible to express with conventional models. This means that agent-based models can provide a means of predicting medium- to long-term changes in downtown areas and may also help people to assess what policies to use to achieve better changes.

The directionality of social simulation using the agent-based approach to shop-around behavior modeling includes approaching the essence of the phenomenon that occurs by developing as simple a model as possible (based on the KISS principle, Keep It Simple, Stupid, proposed by Axelrod), and aiming for a decision-making tool to contribute to the real world and to point to a model with high fidelity for the real world (Kita 2007). Although both directions are possible for shop-around behavior modeling, most research is oriented toward models, with high fidelity models conveniently constructed from the standpoint of social engineering. In this study, a model was constructed in accord with these directions.

7.1.4 Characteristics of the Model

Based on the above, this model is a non-aggregated model that uses the agent-based approach to model shop-around behavior. In addition, it is suitable for analyzing the accumulated effects of diversified consumer behavior because it uses a bottom-up approach that reproduces the individual behavioral patterns of pedestrians at the micro-level and expresses macro-level social phenomena as accumulated results.

The model has a focus on both planned and improvised action. In terms of planning, a formula is used to solve the issue of bounded rational constraints to realize a scheduling function at the agent level for efficiently visiting the shops necessary to achieve the purpose within a limited time period. Meanwhile, as for deviations from the plan and the ensuing revisions to the plan, the model was developed step by step for each hierarchy after hierarchical decomposition (see below for more detail). Furthermore, the learning functions of visitors are assigned to agents by setting parameters that express a unique preference for facilities and streets to each agent and allowing them to make revisions according to the behavior results.

7.2 Concept Category of Shop-Around Behavior

The concept of shop-around behavior can be categorized by focusing on and considering two characteristics: planned action and improvised action. Planned action refers to actions whereby visitors plan which facilities they will visit and when prior to visiting the commercial districts. Unplanned action occurs due to various factors imposed on visitors by the environment and is a course of action referred to hereafter as "improvised action."

Table 7.1 Functions of pedestrian shop-around behaviors

	Phase	Process		Meaning
		Shop-visit	Path	
Planned action	Plan phase	Visit planning	Path planning	Schedule
	Do phase	Following planned visit	Following planned path	Digest
Improvised action	Accident phase	Impulse visit	Detour action	Insert
	Revise phase	Alternative visit	Path adjustment	Re-schedule

With this in mind, I have attempted to break down shop-around behavior into two axes: time sequence and processing target. I believe that shop-around behavior can be categorized into four phases on the time sequence axis: (a) the Plan Phase, in which a plan is formed before a commercial district is visited; (b) the Do Phase, in which the plan is digested; (c) the Accident Phase, in which unplanned action is inserted due to external factors; and (d) the Revise Phase, in which the plan is revised based on plan execution results and unplanned action. Furthermore, I believe that the processing target can be categorized into two points: "visiting facilities" and "route." In this breakdown, the (a) Plan Phase and (b) Do Phase are planned actions, and the (c) Accident Phase and (d) Revise Phase are improvised actions. Table 7.1 shows the functional characteristics of shop-around actions broken down in this way.

7.2.1 *Consideration of the Plan Phase*

In the Plan Phase, a plan of action for the shop around activity is drafted in advance. Facilities to be visited are chosen according to the purpose of the visit and the time range for shopping around (time budget), and the visits are arranged in order to make it somewhat efficient. At the same time, the route is chosen according to efficiency and personal preferences. This includes both the decision to visit commercial districts and the choice of commercial district to visit.

7.2.2 *Consideration of the Do Phase*

The Do Phase is the implementation of the Plan Phase. In other words, after planning is completed, the visitor goes to the commercial district, walks according to the planned route, and visits the facilities they have decided on. If the purpose cannot be achieved as planned, it will be handled in the Revise Phase described below. In addition, this phase includes the choice of returning home when it is determined that the limit on the time budget is drawing to a close or when the purpose has been achieved.

7.2.3 *Consideration of the Revise Phase*

In this phase, any necessary revisions are made to the plan. This takes place in instances wherein the visitor could not behave according to the developed plan or in instances wherein the purpose could not be achieved, meaning that a second attempt must be made. Specifically, alternative visits occur in which the visitor selects a facility from among the facilities of which the visitor is aware where the objective may be achieved and the visit order is changed within the time budget as needed. The walking route is also revised accordingly.

7.2.4 *Consideration of the Accident Phase*

Unlike the previous three phases, the Accident Phase is a phase in which higher intellectual functions are handled, as a deviation from the plan is made for the sake of the visitor's own intentions, preferences, or desire to search. When a visitor visits unplanned facilities or facilities that they did not know exist, it is largely due to this action, which has an economic effect brought about by the accumulation of facilities in commercial districts. Below, I consider impulse visits that occur when visiting facilities and detour action that occurs in route choice, which are included in this phase.

1. A study of impulse visits
   An impulse visit is a visit that occurs when something motivates a pedestrian to visit a facility without having chosen an errand in advance. The impulse visit is an improvised visit to this facility in line with that motivation. This study categorizes the action according to the type of motivation to visit a facility, as shown in Table 7.2. Of these, it is possible that the Type-II latent motivation is included in a wider sense in the categories of planned action/alternative visits based on the idea that this motivation existed prior to the formulation of a plan. Therefore, Type-I

Table 7.2 Classification of motivations for impulse visits

Type-I	Impulse visit generated to recognize the target shop by passing or by seeing its signboards or posters	
	Ia	Target shop is unknown before recognition
	Ib	Target shop is already known before recognition
Type-II	Visit in the event that motive was generated after the plan was drawn up	
	IIa	Generated motive specifies the target facility
	IIb	Generated motive does not specify the target facility

should hereafter be taken to refer to impulse visits in a narrow sense, and this is what "impulse visit" will refer to unless stated otherwise.

2. Study of detour action

Regarding the actual visitor detour action, it is difficult to make an accurate assessment. In existing research, there are many examples in which the degree of separation between the actual walking route and the shortest possible route is given as an index. This method is not able to distinguish between (a) cases in which the shortest route is not chosen during planning and (b) cases in which an impulsive change of route is made during the shop-around. Using the categories in Table 7.2, (a) occurs during planned route creation during the Plan Phase, and (b) occurs as a detour action during the Accident Phase, which makes it the target here.

As with impulse visits, motivations can be categorized as cases in which the route is changed impulsively in recognition of the target route and cases in which the change to the route is made intentionally due to demands that arise after planning; however, in this study, only the former is handled as a detour action, per impulse visits.

7.3 Composition of Shop-Around Behavior Simulation Model

7.3.1 The Concept Behind This Model

The Agent-based Simulation for Shop-around Agent (ASSA) model is comprised of three models: (1) the pedestrian agent model; (2) the hierarchical commercial district model; and (3) the stay-at-home model. (1) represents the pedestrian and his/her behavior, which is the most important part of this model. (2) is the part that represents the commercial district visited by the pedestrian agent. In line with the purpose of this section, the idea in mind is that of a downtown area. (3) is the abstract space where various daily activities are conducted by the agent outside of the visited commercial district. The behavior that refers to the preparation of shop-around behavior, such as the formation of the agent's intention to visit the commercial district, planning, and revision of knowledge bases after the visit, function while the agent is in this space.

7.3.2 The Pedestrian Agent Model

The agent behavior model when assuming a unipolar commercial district is composed of three models: (1) the planning model while at home; (2) the shop-around model inside the commercial district; and (3) the travel model between home/

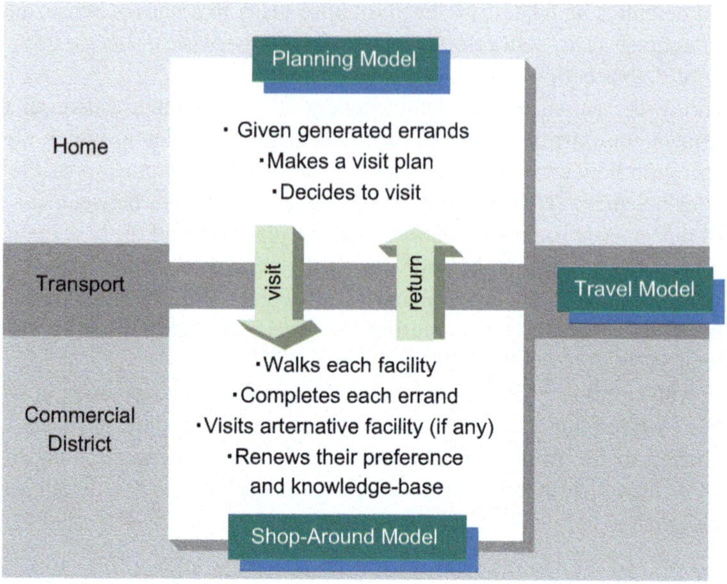

Fig. 7.1 Agent behavior model

commercial districts. With (1), various errands and time budgets are given to each agent, and the agent makes decisions about visiting the commercial district in this regard. An agent that chooses to visit the commercial district drafts the time of the visit and an action plan for the visit to the commercial district based on the given limitations and his/her knowledge base. With (2), the agent that has visited the commercial district based on the plan formed in (1) walks around the facilities for the purpose of achieving the errand(s). In the case of a failure to achieve said errands, an improvised action is produced in which improvised changes are made to the plan. In addition, improvised revisions are made to the agent's knowledge base because of the results of this shop-around behavior, which is utilized in future commercial district visits. (3) connects (1) and (2) and is the movement model that represents the round-trip between the agent's home and the commercial district. Figure 7.1 shows the agent behavior model concepts stated above.

7.3.2.1 ASSA Development in Stages

The development of ASSA is divided into three stages based on the implemented level of functionality. In Stage 1, the Plan Phase, Do Phase, and Revise Phase were equipped as a hierarchical disassembly of the shop-around behavior stated above. These phases form the basis of the model where the plan is drafted, implemented and changed depending on its success. Hereafter, the ASSA model equipped with these functions is referred to as ASSA1.0.

In Stage 2, impulse visit functions were added to ASSA1.0, which is known as ASSA2.0.

In Stage 3, further expansions were added for planned route generation functions and detour action functions.

7.3.2.2 Development of ASSA1.0

Assumptions Made During Construction

The following assumptions were introduced during the development of ASSA1.0.

1. *The city model is a unipolar city, and there are no influential commercial districts outside the represented district.*
   Actual commercial districts often have either competitive or reciprocal relationships with other nearby commercial districts, and it is important to choose a specific commercial district during decision-making for commercial district visits. However, when modeling multiple commercial districts, a significant amount of time is needed for calculations for micro level simulations, which makes such simulations impractical. Furthermore, if commercial district visit decision-making and the evaluation of visit results are excluded, the existence of other commercial districts is not necessary. Therefore, to simplify this simulation, there is only one commercial district in this city model.
2. *There is no direct interaction between agents.*
   Interaction between agents is a peculiar characteristic of agent simulations that cannot be represented in other simulation formats. The interaction in shop-around behavior may include interaction among the family of the agent and group-behavior including acquaintances and friends in the commercial district. However, all such interaction was omitted from this the representation of pedestrian behavior in this model because it would only present a small obstacle if added at a later date.
3. *All agents are familiar with the routes and facilities.*
   It would be preferable that knowledge of the routes and facilities is increased/revised according to the knowledge of the agent, but to include this would carry the risk of complicating the evaluation of the basic plan behavior model. Therefore, it was omitted to more accurately represent behavior at this stage.
4. *Route selection is made only on the basis of route length.*
   During plan drafting and when walking, it is thought that pedestrians select routes that are close to the shortest route by evaluating the street environment (Takeuchi 1977; Kikuchi et al. 2002). However, in ASSA1.0, to confirm that route selection is made reliably, the evaluation of street environments has been omitted and only route length is considered. These assumptions are omitted from ASSA3.0, as will be discussed in more detail later.

Visit Preparation Model

Figure 7.2 shows the visit preparation model, which comprises the following modules:

1. Errand Generation Module. This generates errands that are to be carried out in the commercial district.
2. Commercial District Visit Decision Module. This makes decisions concerning the visit according to the number of errands.
3. Time Budget Generation Module. This generates a time budget that consists of the visit day and the possible time-bands for the visit.
4. Plan-Making Module. This generates a plan to efficiently complete the accumulated errands within the constraints of the time budget.

Of these, the functions of visit planning and path planning in the Plan Phase are achieved in (4). Below, the details of each module are given.

1. *Errand generation module*
   The errand generation module provides the errands that need to be completed in the commercial district by the agent. The number of errands (*n*) per day given to each agent is based on a Poisson probability, based on a deduction of the study results. Each errand has a total of four properties: classification (equivalent to facility classification); corresponding facility; time required for achievement; and limitations. The method for determining each property is as follows:
   (a) Classification is determined by the preference calculated for each agent based on the errand classification selection matrix for each agent's attributes.
   (b) The corresponding facility is selected by searching the agent's knowledge base of the facilities classified above. As for the search and the selection,

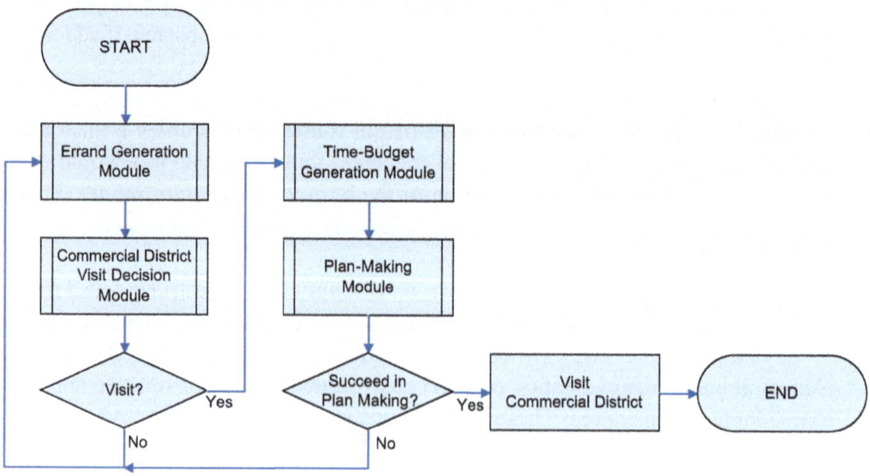

Fig. 7.2 Flowchart of the visit preparation model

using an ε-Greedy algorithm, a facility is selected from the corresponding facilities that has the highest probability $(1-\varepsilon)$ of screening variable (w), and the corresponding facilities with the probability ε are selected at random (however, facilities with negative w are excluded).

(c) Time required for achievement is produced based on a normal distribution, based on the average μ/standard deviation ρ given to each errand classification.

(d) Limitations are randomly assigned from among the six categories below.

- Type-1 (Complete closure-type):
  Fixed limitations at start of errand, floating-type limitations at end.
- Type-2 (Fixed-at-start type):
  Fixed limitations at start of errand.
- Type-3 (Closed at both ends-type):
  Floating-type limitations at the start and end of the errand.
- Type-4 (Closed at start-type):
  Floating-type limitations at the start of the errand.
- Type-5 (Closed at end-type):
  Floating-type limitations at the end of the errand.
- Type-6 (Complete open-type):
  No limitations at either the start or end of the errand.

  The fixed-type limitations and floating-type limitations are defined below.

  - Fixed-type limitations:
    Errands that must start at the exact time of the limitation (e.g., concert, movie, hospital appointment, etc.)
  - Floating-type limitations:
    Errands that can be performed freely within the period of the limitation (e.g., general shop opening/closing hours)

2. *Commercial district visit decision module*
   When an agent's number of errands is x, the probability (p) that the agent will enter a commercial district is in accord with the accumulated Poisson distribution shown below.

$$p = \sum_{i=1}^{x} \lambda v^x e^{-\lambda v} / x! \quad (7.1)$$

However, λ_v is determined as follows by the attributes of the agent.

$$\lambda v = \lambda g \times \lambda ad \times \lambda ag \times \lambda w \times \alpha \quad (7.2)$$

α is the constant given when setting the commercial district model. As an example of each value, Table 7.3 shows the findings based on the analysis of the data of the Osu district shop-around behavior study discussed later.

Table 7.3 Estimated values to calculate λ_v

	Attribute of agent	Value
Gender	Male	0.926
	Female	1.108
Address	Near	0.828
	Middle	1.510
	Far	1.884
Age	Less than 30	0.857
	30–49	1.184
	50 or more	0.920
Work	Company employee	1.107
	Self-employed	0.481
	Housewife	1.782
	Student	0.885

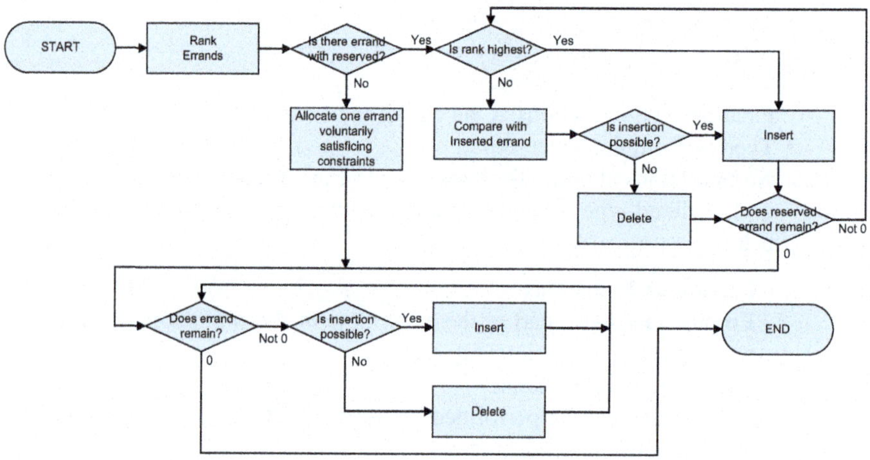

Fig. 7.3 Flowchart of the planning module

3. *Time budget generation module*
   The time budget is given to an agent by producing a visit schedule and probable action time based on a normal distribution probability, after determining the agent's commercial district visit based on the commercial district visit decision module.
4. *Planning module*
   In the time budget given in the time budget generation module, the action plan is formulated to achieve the accumulated errands as efficiently as possible in one round-trip between home and the commercial district. At this time, the following algorithm is used to obtain the degree of satisfaction for the limitations attached to the errands (see Fig. 7.3).

1. A random order is given to all errands.
2a. If there are errands with fixed-type limitations, they are fixed first.
2b. If there are no errands with floating-type limitations, only the errands of a high order in "1." are fixed.
3. The errands without fixed-type limitations that remain after completing "2." are allotted to the surplus areas of the time budget in line with the order of "1."
4. Allotment is complete when all errands have been allotted, or when the remaining errands can no longer be allotted, and they are arranged as a temporary plan.
5. Steps "1." to "4." Are repeated *n* times.
6. The produced temporary plans are evaluated in alphabetical order using the two variables; one is plan length (number of errands included in the plan) and the other is planned action time length. The plan with the highest value is selected.
7. The non-selected errands are eliminated.
8. The arrival point that is closest to the errands at the top of the plan is selected as the point of entry to the commercial district.

Figure 7.4 shows the draft of an action plan using the above algorithm.

Commercial District Shop-Around Model

The commercial district shop-around model comprises the following four modules.

1. Trip to Facility Module. This expresses the moves each agent in the commercial district makes by applying the Dijkstra method.
2. Errand Achievement Module. This determines the success or failure of an errand that the agent attempts.
3. Alternative Visit Action Module. If the agent fails to complete the errand, a decision is made whether to carry on with the errand, search for an alternative facility, or modify the plan.
4. Post-action Processing Module. After completion of the errand(s), a decision is made concerning the return home, and the preferences are updated.

Of these, **Following the planned path** in the Do Phase is completed in (1), and the function of **Following planned visit** is completed in (2). In (3), the functions of **Alternative visits** and **Path adjustment** from the Revise Phase are completed. The details of each module are discussed below (Fig. 7.5).

1 *Trip to facility module*
This module represents the trip by the agent inside the commercial district. After arriving at the commercial district, each agent reads the properties of the facility corresponding to the first errand at the top of the action plan and establishes the target area as the destination in the corresponding facilities. The shortest route to the target area is then calculated using the Dijkstra method, and the agent moves within the commercial district in line with this. After arriving at the target area, if

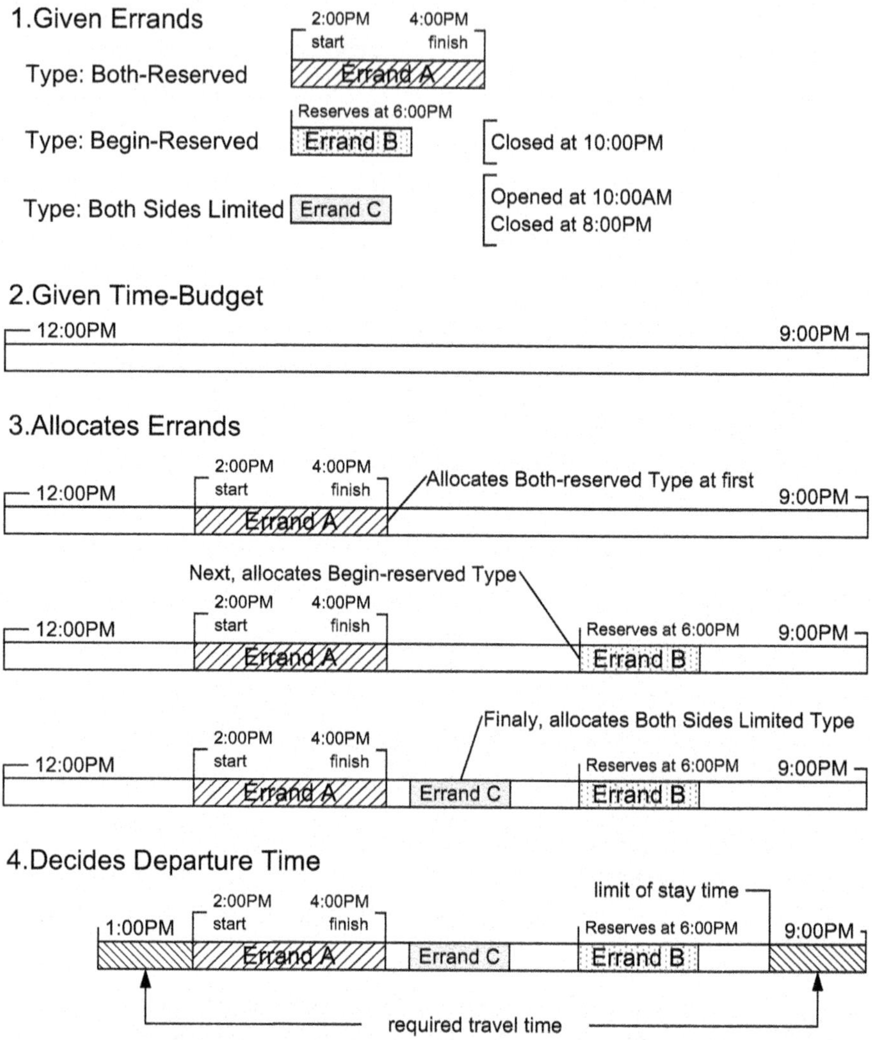

Fig. 7.4 Example of formulation of the action plan

the time limitations are satisfied, the implementation of the errand achievement module begins immediately; if the time limitations are not satisfied, the agent remains on standby at the appropriate point.

2. *Errand achievement module*
   This module determines the achievement of E0 when the agent acts to complete the errands E0 in the store variety in the commercial district. At the point after the time required for achievement has passed after starting to complete the errands, the achievement of the errands E0 is determined by the probability p_0 $(0 \leq p \leq 1)$ based on a uniform random number. After achievement, the corresponding

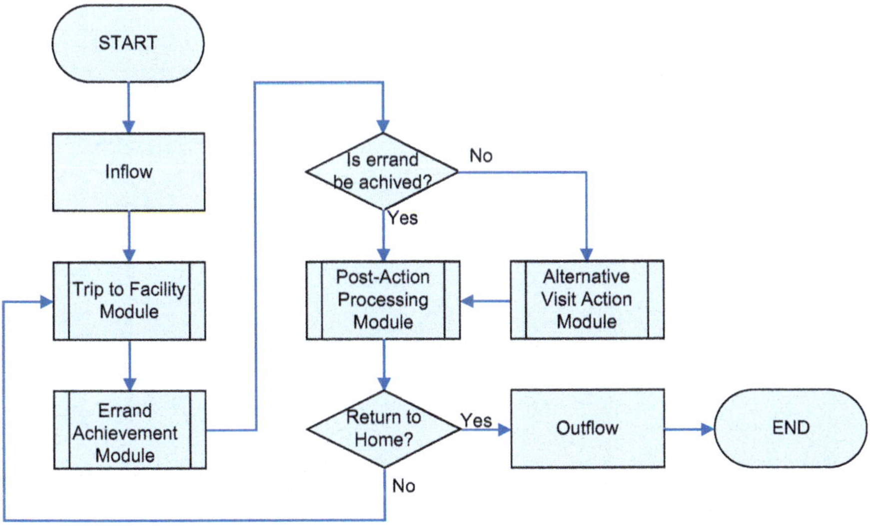

Fig. 7.5 Flowchart of shop-around model

facility properties for the errands E1 are found and the target area is set. Meanwhile, if E0 has not been achieved, the following alternative visit is produced.

3. *Alternative visit module*
   If the errands E0 are not achieved in the errand achievement module, a decision is made as to whether to maintain the objective of achieving E0. A search is conducted for alternative facilities for that purpose, and a new action plan is constructed. At this time, in the search for alternative facilities, the corresponding facilities are selected at random with a probability of ε using an ε-Greedy method, and alternative facilities with a performance function are selected with the highest value of probability $(1-\varepsilon)$, as shown below.

$$f(l,\beta) = w \times \beta + (1-\beta)/(l+1) \tag{7.3}$$

$$\beta = t_{\mathrm{R}}/t_{\mathrm{A}} \tag{7.4}$$

However,

w: Preferential value of corresponding classification of facility
l: Distance from current location to alternative facilities (calculated using Dijkstra method)
t_{R}: Possible duration of stay remaining in commercial district
t_{A}: Possible duration of stay in commercial district

The Plan-Making module is retrieved after selecting alternative facilities and an action plan is reorganized. In addition, if β from Eq. (7.4) is less than 0.1, or if there are no fixed-type limitations on errand E0, an alternative search is conducted toward achieving the next errand.

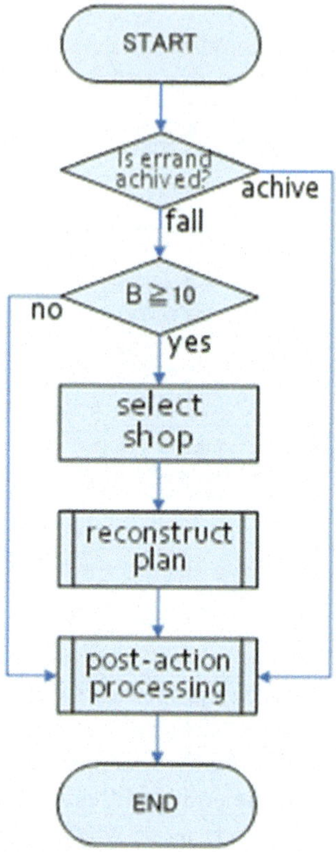

Fig. 7.6 Flowchart of errand achievement module/ alternative visit module

Figure 7.6 shows the flowcharts for (2) and (3) from above.

4. *Post-action processing module*
   After deciding on errand achievement, this module describes the processes performed by the agent. Specifically, two processes are performed: (a) fluctuation in preferential variable w; and (b) the decision to return home.

   (a) Fluctuation in preferential variable
   The preferential variable w fluctuates depending on errand achievement. The value of fluctuation f follows a normal distribution with a mean of $\mu = 3$ and a standard deviation of $\rho = 1$.
   The initial value of the preferential variable is set in line with a normal distribution of $\mu = 5$/standard deviation of $\rho = 2$ for all facilities.

   (b) Decision to return home
   When the agent has achieved all errands or when the time limit for action is reached, he/she decides to return home. At that time, the departure point is set as the nearest station to the current location for train users and the car park that was used when visiting the commercial district by automobile users.
   The flow of the above processes is shown in Fig. 7.7.

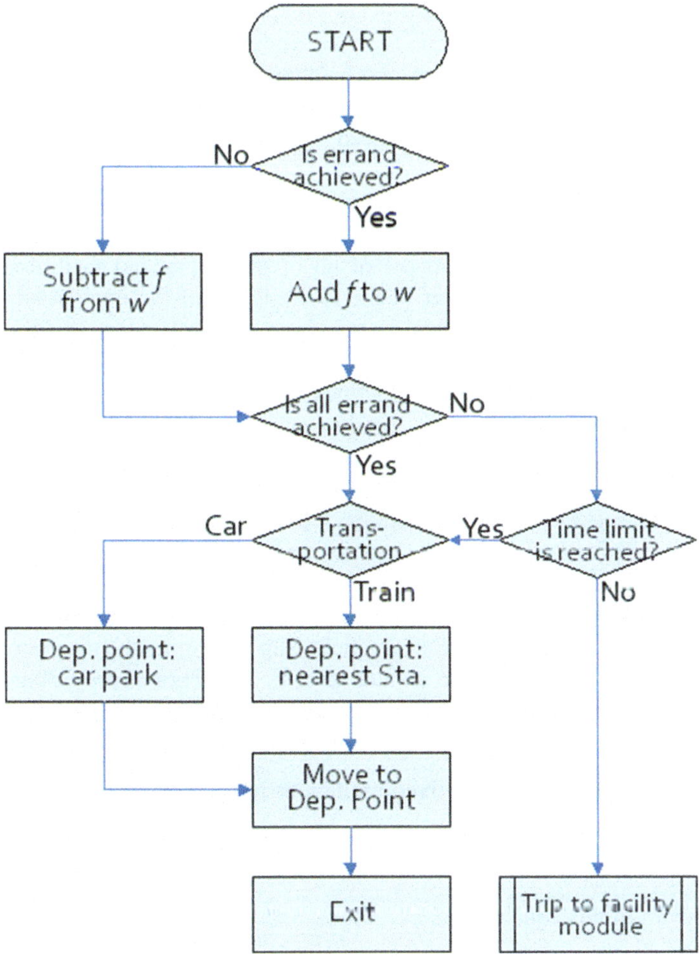

Fig. 7.7 Post-action processing module flowchart

Model of Movement Between Home and the Commercial District

This model represents the round-trip between home and the commercial district. In the case of visiting the commercial district, after the time required for transportation *T* has elapsed after leaving the home, the agent arrives at the entry point of the commercial district as set in the Plan-Making model. Meanwhile, when leaving the commercial district, the agent moves to the departure point set in the post-action processing model and returns home after the time required *T* has elapsed. The time required from the home to the commercial district is set according to the residential location of the agent.

7.3.2.3 Development of ASSA2.0

Regarding ASSA1.0, as discussed above, after confirming that the models were expressing the Plan Phase, Do Phase, and Revise Phase function as intended, a model that realizes the more advanced function of the Accident Phase was constructed.

During construction, a model that represents impulse visits in the Accident Phase was first constructed. However, in ASSA1.0, the assumption was that all of the agents mentioned above were aware of all of the routes and facilities, and this assumption was also carried over in the construction of the impulse visit model; thus, impulse visit Type-Ia in Table 7.2 does not necessarily occur, and the modeled target was Type-Ib only, that is, only the action of visiting known facilities impulsively. The ASSA that includes this function is called ASSA2.0. Below, the model of impulse visits is described.

Each agent, when staying on the streets in the commercial district, received a facility information value (I_{FC}) from the facilities on the streets. I_{FC} is an imagined value that presents the various attractions of the facilities. Next, the received I_{FC} is corrected by the facility classification preferential value (W_{FT}) characteristic to each agent. At this time, the agent will pick up only the received I_{FC} with n information from among those with a high value, which is reflected in their actions. Then, the average value (I_{AVE}) of the I_{FC} from the picked-up n becomes the motivator, and the surplus time ratio (γ) that can be divided to carry out actions becomes an opportunity; these are referred to in order to make a decision as to whether or not to make an impulsive visit to the facilities. At this time, γ is defined as follows.

(a) In the case that there are no fixed limitations to the following errands

$$\gamma = t_{\mathrm{R}}/t_{\mathrm{A}} \tag{7.5}$$

However, in this equation

t_{R}: Remaining stay time
t_{A}: Total stay time

(b) In the case that there are no fixed limitations to the following errands

$$\gamma = \delta\left(1 - \frac{t_{\mathrm{L}}}{t_{\mathrm{C}} - t_{\mathrm{N}}}\right) \tag{7.6}$$

However, in this equation

δ: Factors determined for each district
t_{L}: Time distance to destination
t_{C}: Time limitations of next errands
t_{N}: Current time

If the decision is made to implement an impulsive facility visit based on the judgment above, with reference to the two variables of the information value of the pickup facility and the preferential W_{FC} value regarding each facility, the agent decides which facility to visit. In addition, by means of this visit, if it is necessary to change the action plan, the plan is replaced using a developed alternative visit function.

The series of algorithms described above is shown in Fig. 7.8.

7.3.2.4 Development of ASSA3.0

By means of the development of ASSA2.0 equipped with impulse visits, a total of four functions were realized, including advanced functions for processing visit facilities. After confirming that it was functioning as intended, as with ASSA1.0, I also engaged in modeling detour action as an advanced function for route processing. As with this function, I also set about installing a planned visit action that provides a deviation from the shortest route. The ASSA that includes these functions is called ASSA3.0. These models are described below.

(a) Planned visit action model
Planned visit actions arise during the formation of the planned route during the Plan Phase. When constructing a model, setting apart assumption "4" in ASSA1.0, this is realized by expanding the Plan-Making module in the visit preparation model. Planned visit actions are divided into cases in which the shortest route cannot be chosen because the pedestrian is unaware of all of the routes, and cases in which a route is chosen that is not the shortest route for reasons of safety, congestion, or preference despite knowing the shortest route. However, as with impulse visit Type-Ia, because of the assumption that all agents are aware of all of the routes and facilities, the former situation does not occur; therefore, in this model, only the section in which an agent intentionally deviates from the shortest route is represented.
To equip this action model, a technique was adopted that gives a different weighting to each agent for each link on the commercial district model network. Depending on this weighting, because of the existence of short routes according to the differing mentalities of each agent regarding geographically shorter routes that are determined materially; although the agent should have drafted an action plan after personally choosing a short route, in fact, deviations were already made from the shortest route during planning. At that time, regarding the weighting given to each link, a random element was added to the total preferential values of each agent for the facilities that exist on each link.

(b) Impulse visit action model
Impulse visit actions correspond to deviations from the planned route during the Accident Phase that are a higher function in the form of a pair with impulse visits. When modeling this action, based on the following processes, the softmax technique (for details of this technique, please refer to Takadama) using a Boltzmann distribution was applied. First, a point that is separate from the actual

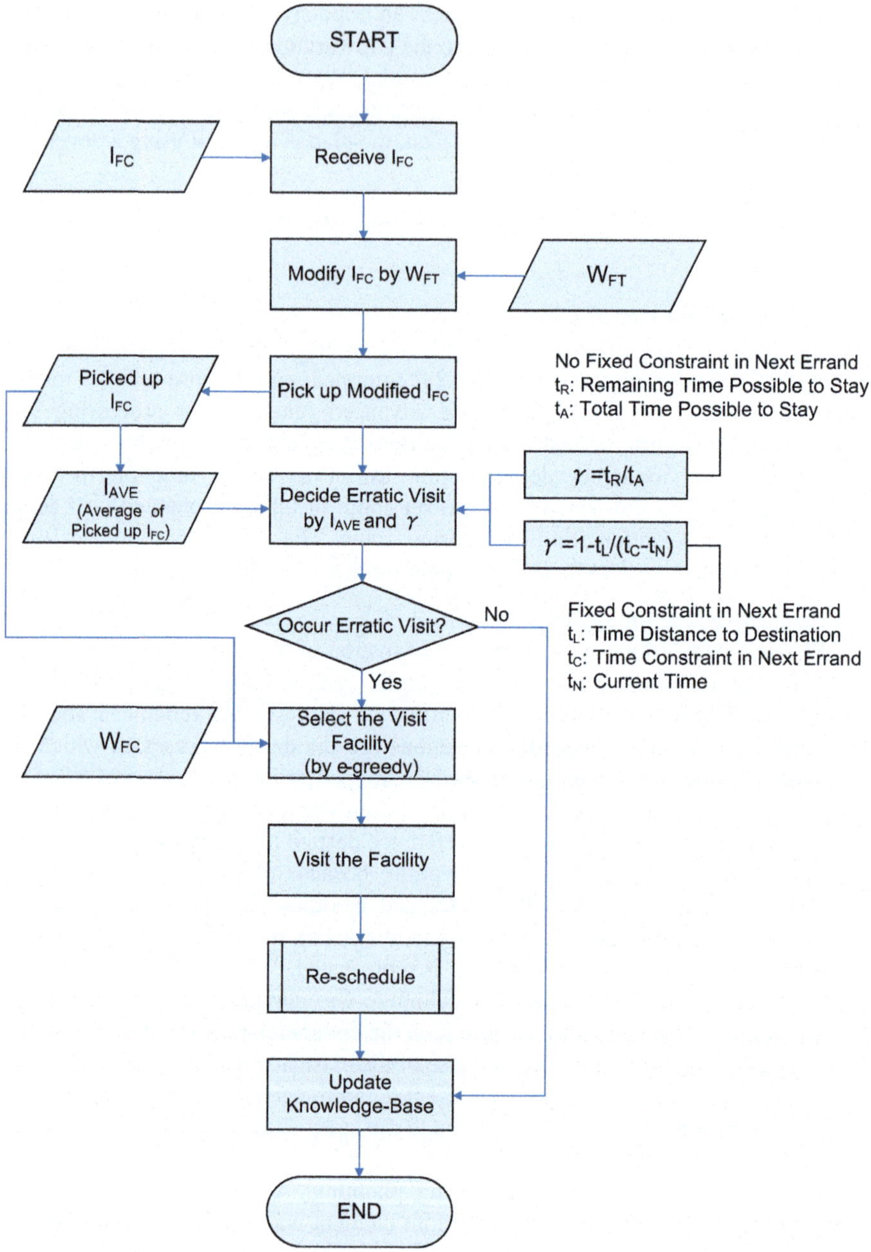

Fig. 7.8 Algorithm of the impulse visit model

purpose points during the commercial district shop-around was randomly selected. The value of this point was then calculated based on the distance and weighting (as used in the plan visit action model) between the current point and each point. Based on this value, the probability of selecting each point was determined using the Boltzmann distribution (see Eq. 7.7) that has given γ (see Eqs. 7.2–7.6) as a temperature parameter. However, when β (see Eqs. 7.2–7.4) is $\beta < 0.1$, no impulse visit actions occurred.

$$p(a|s) = \frac{e^{Q(s,a)/\gamma}}{\sum_{a_i \in A} e^{Q(s,a_i)/\gamma}} \tag{7.7}$$

Regardless of planning/shop-around, no particular addition was made to the function of facility visits for the installation of visit actions. This is because facility visits during visit actions occur automatically due to the existence of the impulse visit model described above.

7.3.3 Hierarchical Commercial District Model

The commercial district spatial model is represented by a three-stage hierarchical structure.

1. Whole district model representing the entire commercial district.
2. Area model representing each area in the commercial district.
3. Routes between areas that connect each area.
4. Model of facilities existing in each area.
5. Routes in areas connecting facilities.

In this model, (1) has the top position, and includes (2) and (3). Equally, (2) includes (4) and (5).

In this model, routes between streets always connect the center of each street. Routes in streets also connect the center of each street with each facility. In other words, the concept of distance is represented by routes between streets and the street itself has no distance. The distance of each street is universally assumed to be 0, so there is no difference in distance between facilities that exist on the same street (Fig. 7.9).

7.3.4 Stay-At-Home Model

Here, "stay-at-home" refers to the abstract space wherein the agent is on standby when not visiting commercial districts. Except for commercial district visits, the daily life of a pedestrian in its entirety is represented in the form of being in a "stay-

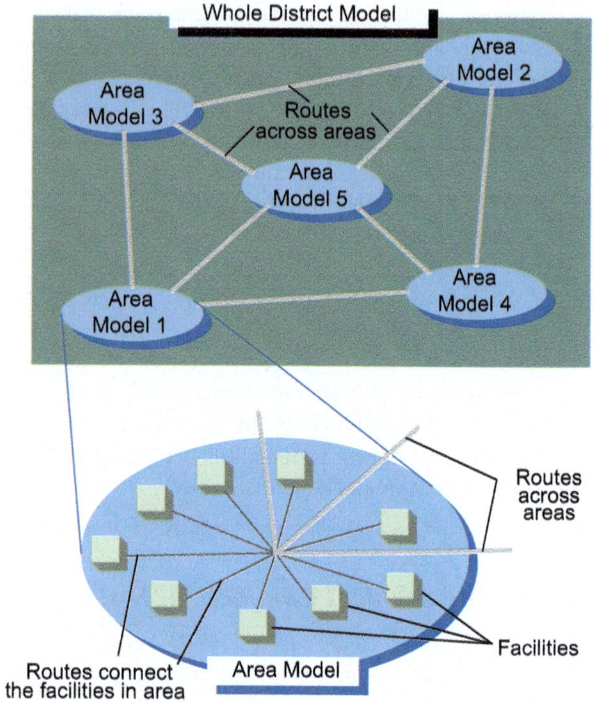

Fig. 7.9 Concept of the 3-layered district spatial model

at-home" model. In addition, the implementation of functions such as commercial district visit decision-making, plan-making, and revisions to the agent's knowledge base after the visit is performed while the agent is in the stay-at-home model.

Only the concept of distance from the commercial district model is given to the stay-at-home model. The distance is either "near," "middle," or "far," calculated as distances of "30 minutes," "1 hour," or "1.5 hours" from the commercial district, respectively. As for other spatial concepts, representation was omitted because such concepts are not required for the purpose of developing this model.

7.4 Construction of a Model Evaluation Framework

7.4.1 Establishment of Evaluation Points

To evaluate the characteristics of the above designed model, the simulation results have been analyzed considering the following five points.

(a) General evaluation

1. Analysis of macro-behavior
2. Analysis of statistics

(b) Individual function evaluation

3. Illustration of the agent's individual behavior
4. Analysis of similarity of the visit sequence
5. Analysis of redundancy

The evaluation method consists of two categories: (a) general evaluation, which can be defined as a broad examination of the overall simulation result; and (b) individual function evaluation, which can be defined as verification of each function of the model. With regard to general evaluation, I evaluate from two points of view: in (1), the accumulated results of agent walking flow lines and facility visits are compared to the data of a pedestrian survey so as to examine agent behavior as a macro-simulation; and in (2) statistical results representative of agent behavior are examined and in the same manner as (1), by comparing the result to the survey data, the characteristics of agent behavior can be clarified.

Individual function evaluation is carried out from three points of view. In (3), which concerns plan-making and implementation, and the occurrence of unplanned actions and any subsequent revision by an agent, evaluation is carried out by tracking the behavior result of each agent, and it is confirmed whether each behavior function is executed as designed. In (4), the similarity between the simulation and the survey is measured. Moreover, as in (2), in (3) the distance walked by agents is compared to the survey data to clarify the results of route selection and unplanned actions, as well as the efficiency of the planned schedule.

Evaluation based on the above five points allows the characteristics and value of the model to be understood and provides a guide for future development. The following sections describe the details of each evaluation method.

7.4.2 Macro-Behavior Analysis

To conduct a comparative analysis of the behavior results of agents, there should be a focus on the walking percentage and the number of facilities visited for each street. Diagrams should be used to carry out comparative analysis. In this research, the walking percentage was obtained by dividing the number of pedestrians on each street by the number of visitors. To obtain the number of facilities visited, the sum of facilities visited within each street is divided by the number of visitors. A higher walking percentage for a street indicates that the street is busy, and a larger number of facilities visited on a street means that the facilities themselves are crowded.

7.4.3 Statistics Analysis

Analysis is carried out by focusing on statistics that represent the behavior results of agents. More specifically, attention is paid to each agent and survey sample value for

(1) the duration of the stay, (2) the number of facilities actually visited, and (3) the number of planned visits to facilities. Tests for differences between means and a chi-square test for a contingency table are conducted. Based on these tests, and by comparison with the actual survey data, characteristics of agent behavior can be clarified.

7.4.4 *Illustration of Agent's Individual Behavior*

The individual behavior results for each agent can be extracted from the data obtained by the simulation, and this can then be expressed graphically on the map. There are two focal points: facilities visited and walking routes. To observe the facilities visited, attention is paid to the following four points: locations; visiting order; planned or unplanned visits; and task success or failure. Based on my observations, I check whether each designed function is working: validity of a plan drawn up by an agent in the Plan Phase; implementation of the plan in the Do Phase; occurrence of impulse visits in the Accident Phase; and occurrence of alternative visits in the Revise Phase. Walking routes are assessed mainly by paying attention to whether agents have deviated from the shortest route. I then check the normal planning and implementation of a planned route, and the occurrence of detours from the planned route.

7.4.5 *Analysis of Similarity in Visit Sequences*

Facility visiting patterns for each sample from the simulation and the survey are compared to verify how closely the following points resemble each other: the facilities visited, visiting order, and the number of facilities visited.

For verification, the concept of Levenshtein distance is introduced. An agent sample and survey sample are corresponded one-to-one, between the visit sequences, the Levenshtein distance is calculated by assigning a cost of 1 for each insertion, deletion and substitution, and this result is used as an indicator to assess to what extent the result of agent behavior has reproduced facility visits by a real pedestrian. To determine the visit sequence of a certain agent, survey samples whose gender and age belong to the same category are selected. From these, the survey sample with the shortest Levenshtein distance is used, while both an average and a variance of the shortest Levenshtein distances as a whole are obtained to conduct variance analysis. The percentage of sample pairs with a Levenshtein distance of 0—meaning the visit sequences are identical—is also used as an indicator.

In this case it is important to note that the Levenshtein distance is determined according to the one-to-one correspondence results of the data and survey sample;

therefore, the average and variance of Levenshtein distance for the whole sample change according to the correspondence determination method. Consequently, to maintain the validity of this indicator, a method to find a solution that is objectively acknowledged as the optimum sample pair must be adopted. For this problem, it was decided to consider the problem as an application of a stable marriage.

If we regard "a man" in this problem as an agent sample, "a woman" as a survey sample, and the "ranking" as the Levenshtein distance from the marriage partner, by matching stable couples with samples, it is possible to find the optimum sample pair. However, there is a plurality of solutions for stable coupling, so a plurality of solutions is calculated by changing the permutation of samples, and the solution that gives the minimum average value of the shortest Levenshtein distance is adopted.

7.4.6 *Redundancy Analysis*

In both the simulation and survey, attention is paid to the degree to which the behavior results and the scheduled plan deviate from the optimum geographical distance solution. By comparing them, behavior and planning characteristics in the model are verified. The similarity analysis of visit sequences in the previous section pays attention to the visiting order and examines the degree of similarity between samples, whereas this analysis focuses on geographical distance and compares the degree of deviation from the optimum route distance. For verification, detour behavior indicators consisting of the three levels below are adopted, as proposed by Arakawa (Arakawa and Kaneda 2002).

7.4.6.1 Level 1 Indicator

This indicator expresses the degree of deviation between the actual distance walked between facilities and the shortest distance between facilities. This indicates redundancy concerning a route between facilities and can be interpreted as a detour.

7.4.6.2 Level 2 Indicator

When the shortest route distance between one planned-to-be-visited facility (hereinafter planned-visit facility) and the next is regarded as a benchmark, this indicator represents how much the shortest route distance between the facilities actually visited deviates. This indicates redundancy concerning an unplanned visit, and when this value is high, it can be interpreted as behavior extended to include unplanned visits to facilities some distance away, in order to fulfill a new task within the district.

7.4.6.3 Level 3 Indicator

When the route distance of the shortest sequence between planned-visit facilities is regarded as a benchmark, this indicator shows how much the shortest route distance of the actual visit sequence between the planned-visit facilities deviates. This indicates redundancy concerning a plan for the sequence of facilities visited, and it can be interpreted that the lower the value, the more efficient the sequence taken.

By using hierarchical decomposition of pedestrian behavior in 2.2, these indicators can be reinterpreted as follows: the Level 1 indicator shows how often a detour from the planned route occurs in the Accident Phase; the Level 2 indicator shows how often impulse visits occur in the Accident Phase, and how often alternative visits and route revisions occur in the Revise Phase; and the Level 3 indicator shows the degree of deviation between the plan made by a pedestrian in the Plan Phase and the optimum solution. Accordingly, using these indicators to compare the results of agent behaviors with that of survey samples allows us to verify the following: characteristics of detours from the planned route in the Level 1 indicator; characteristics in the combination of erratic and alternative visits in the Level 2 indicator; and the efficiency of a plan in the Level 3 indicator.

References

Arakawa S, Kaneda T (2002) An analysis on redundancy of ‘Shop-Around’ behavior in Nagoya CBD. J Architect Plann Environ Eng (Trans AIJ) 556:227–233. (in Japanese)

Borgers A, Timmermans HA (1986) A model of pedestrian route choice and demand for retail facilities within inner-city shopping areas. Geogr Anal 18:115–128

Kaneda T, Yokoi S, Takahashi S (2001) Development of pedestrian shop-around behavior model considering transition between facilities. Simul Gaming 11:17–23. (in Japanese)

Kikuchi A, Ohgai A, Kitano M (2002) Route selection behavior analysis for the development of pedestrian behavior model in urban area (part 1). AIJ Conference Proceedings 2002 F-1, 625–626

Kita H (2007) Agent-based approaches to social simulation. IPSJ SIG Notes 2007(120):23–28. (in Japanese)

Kurose S, Borgers A, Timmermans HA (2001) Classifying pedestrian shopping behavior according to implied heuristic choice rules. Environ Plann B 28:405–418

Mibuta Y, Ohgai A, Okeno S (2000a) Modeling the pedestrian movement using an agent-based model: part 1. In: Summaries of technical papers of annual meeting of Architectural Institute of Japan, 2000-9, pp 827–828. (in Japanese)

Mibuta Y, Ohgai A, Okeno S (2000b) Modeling the pedestrian movement using an agent-based model: part 1. In: Summaries of technical papers of annual meeting of Architectural Institute of Japan, 2000-9, pp 829–830. (in Japanese)

O’Kelly M (1981) A model of the demand for retail facilities, incorporating multistop, multipurpose trips. Geogr Anal 13(2):134–148

Saito S (2000) Survey report on migration behavior in Fukuoka City Center. Fukuoka Institute of Urban Science. (in Japanese)

Saito S, Ishibashi K (1992) Forecasting Consumers' shop-around Behaviors in an agglomerated within a City Center retail environment after its redevelopments using Markov chain model with covariates. J City Plann Inst Jpn 27:439–444. (in Japanese)

Sakamoto T (1984) An absorbing Markov chain model for estimating Consumers' shop-around effect on shopping districts. J City Plann Inst Jpn 19:289–294. (in Japanese)

Takeuchi D (1977) A study on the propensity of pedestrians to choose a route. Proc JSCE 1977(259):91–101

Yuzawa A, Suda H, Nisikawa K (1993) Development of The Comercial District Evaluation Technique Based on Excursion Behavior in a City Center. J JSCE 458(IV-18):73–80. (in Japanese)

Chapter 8
Policy Simulation Trials of the Shop-Around Agent Model

Takumi Yoshida and Toshiyuki Kaneda

Abstract Since 2005, the authors have been working on a research and development project: the Agent Simulator of Shop-Around (ASSA). ASSA is mainly based on a cognitive scientific approach and the layer deconstruction of downtown visitor behavior through data analysis (including redundancy analysis and a series of previously conducted shop-around surveys). ASSA is activity based and has dynamic scheduling, but the agents are adaptive and behave with bounded rationality and learning. This study's authors have previously reported several findings from their research experiences, and this work is the final phase of the ASSA project. The authors have analyzed the 2003 Osu survey data using ASSA ver.3 and visually verified and validated it. This study addresses the results of three different scenario policy simulations to both examine the potential of ASSA as a policy simulator and explore revised strategies of the overall ASSA project. It is concluded that ASSA has the potential performance for ensuring effective policy analysis.

8.1 Introduction

The shop-around behavior model is also known as the Multi-Purpose Multi-Stop (MPMS) model. Since the 1980s, the model has been developed and studied in fields including geography and urban planning, not only for its practical interest, e.g., downtown regeneration and town center management, but also for its theoretical interest in the field of spatial analysis (Kelly 1981; Borgers and Timmermans 1986).

In the 1990s, by applying a combination of the logit model (which enables data-fitting based on utility maximization approximation) and the Markov-chain

T. Yoshida (✉)
Bureau of Housing and City Planning, Nagoya City Office, Nagoya, Japan
e-mail: t.yoshida.dr@city.nagoya.lg.jp

T. Kaneda
Graduate School of Engineering, Nagoya Institute of Technology, Gokiso, Showa, Nagoya, Japan
e-mail: kaneda@nitech.ac.jp

T. Kaneda (ed.), *Downtown Dynamics*, Agent-Based Social Systems 16,
https://doi.org/10.1007/978-4-431-54901-7_8

assumption (which makes up transition probability), origin–destination matrices and the layered learning practice model were established in transportation planning.

Albatross (Arentze et al. 2001) was formulated as a rule-based system that guarantees data-fitting by employing a data-mining tool to generate heuristic rules (binary tree). Aurora (Arentze et al. 2005) was formulated as a utility-based theoretical model that generates a schedule by combining each activity (errand) with nonlinear S-shape utility and employing genetic algorithms; in addition, the model conducts rescheduling in response to an unexpected event (such as congestion).

However, when compared with a daily activity-travel model with the same MPMS structure, the shop-around behavior model's characteristics include human-like planned actions, and it also incorporates the improvised actions of downtown visitors; therefore, as a research theme, the shop-around behavior model faces a higher degree of difficulty. Consequently, few studies are available (Kurose and Borgers 2001; Dijkstra et al. 2009).

The authors' investigations reveal that visitors frequently change their planned and improvised actions. Therefore, based on the data analyses results, a newly devised ASSA (Agent Simulation of Shop-Around, Yoshida and Kaneda 2007, 2012; Kaneda and Yoshida 2012) has been created. ASSA is an activity-based model that addresses agents' spatial behaviors while on shopping trips in downtown areas. In the ASSA model, each agent makes and remakes their own shopping schedule based on time constraints and shopping preferences; they choose alternative trips when errands are not successful, and they make impulse stops at shops and undertake detour actions when time allows. A chain of such activities on 1 day affects an agent's plans for their next visit—and so on. In this context, the agent is naturalistic and wise, behaving rationally in the short term and intellectually in the long term by devising and implementing planning (dynamic scheduling/rescheduling), learning, and adaptation functions.

For this model, a framework of verification and validation has been prepared and conducted, which includes analyses of aggregated behavior, statistics, similarity of visit sequence, and so on (see Yoshida 2010; Yoshida and Kaneda 2012; Chap. 7). This study introduces the results of the verification and validation through intuitive visual presentations rather than by simply presenting the statistical results.

First, this paper refers to existing research and briefly explains the features of the ASSA model with a particular focus on the decomposition of the shop-around behaviors and the system components. The final version (ASSA ver.3) undertakes a dynamic simulation of naturalistic and intelligent shopper behavior. Then, its verification and validation is discussed through the simulated results of the Osu shopping district case. Finally, the study conducts three policy simulations through different scenarios to explore the potential for developing the ASSA ver.3 model into a supporting tool for application in practical planning.

8.2 ASSA: System Design and Components

The first feature of the downtown visitor's behavior to be modeled was the function of time allocation or the scheduling of shop visits under a given time constraint (time budget). Implementing the scheduling function in itself was the antithesis of the Markov model (even if it does emphasize time constraint) and was an application of the agent's intelligent planning function. The concept of shop-around behavior has been explained in cognitive science (Hayes-Roth and Hayes-Roth 1979). The second feature was dynamic updating as part of the agent's behavior, which included rescheduling; this also relates to intellectual functions like adaptation and learning. The third feature of agent behavior to be developed was data-fitting oriented evaluation.

Next, the shop-around behavior in downtown Nagoya was investigated using results from 12 different surveys. The survey asked visitors about their walking routes, the shops visited, and whether these visits were planned in advance. According to the redundancy analysis, the shop-around behavior within a given survey area (including the routes taken and the distance traveled) was found not to be necessarily optimized (Arakawa and Kaneda 2002).

Consequently, it was decided to break down the shop-around behavior into planned and improvised action. Improvised action involves alternative visits and impulse actions (Table 8.1). Planned action is an action performed according to a schedule. Alternative visit action (which is defined as an improvised action) is the action of visiting a shop that was not part of the original plan because the planned errand was not completed in the shop originally visited. This concept was not specifically differentiated in the Markov-type shop-around behavior model.

The basic structure of the ASSA model was designed as a component consisting of several modules, as displayed below in Fig. 8.1. The merits of such modular systems are as a result of piecemeal updating. ASSA ver.1 considers only planned and alternative actions (task completion/failure is represented as a fixed probability for each errand). The rescheduling module is employed in the case of a failure to

Table 8.1 Layer decomposition of shop-around behaviors

Agent's action		Information processing		On schedule plan
		On shop visit	On path	
Planning and decision action		Planning a visit order	Planning path	Forming
Planned action		Visiting planned shop in turn	Following planned path	Keeping
Improvised action	Alternative-visit action	Choosing same category's shop when failed errand	Modifying path	Inserting
	Impluse action	*Impulse-visit (stop) action*: dropping in non-planned shop if time available	*Impulse-detour action*: walking non-shortest path if time available	Re-forming

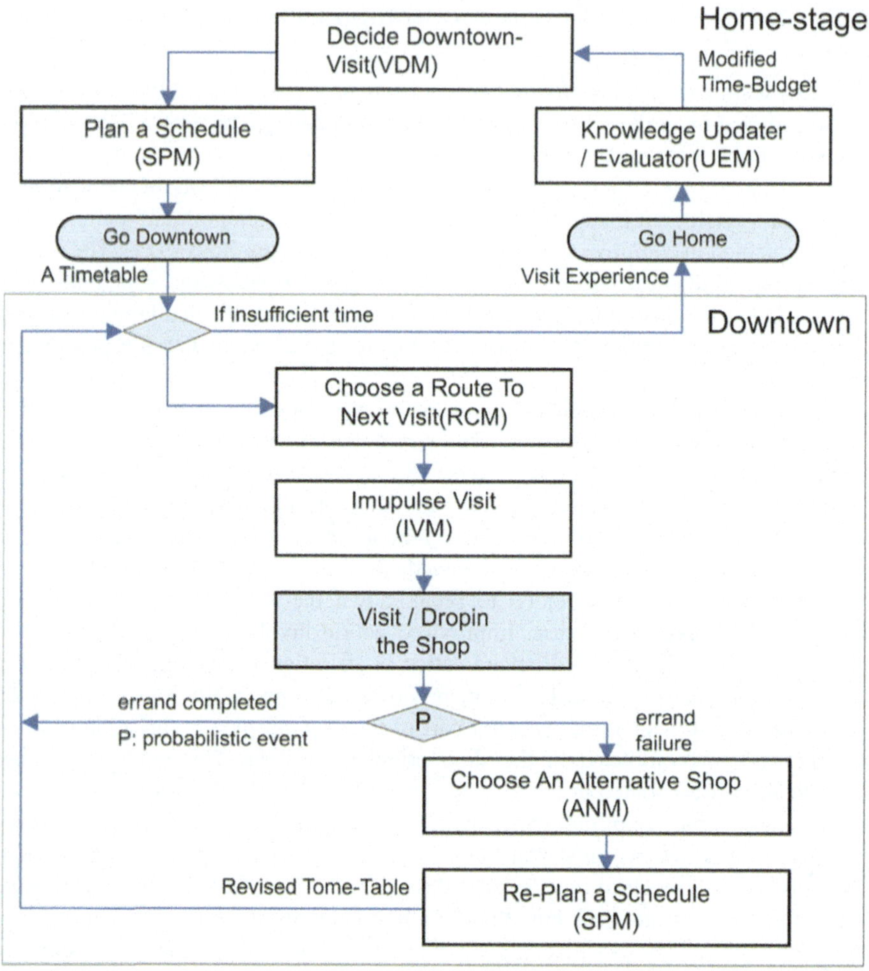

Fig. 8.1 Module components of ASSA

complete an errand. Moreover, the tight assumption presupposes that all path planning is always optimized by Dijkstra's algorithm.

ASSA ver.2 implements the concept of impulse visit action, which is probabilistically determined depending on the availability of adequate time, the street appeal, the sum of the shop preferences along the street, and the time remaining by application of the e-greedy method (for details, see Chap. 7).

ASSA ver.3 adds two kinds of detour actions: one is the relaxation of the optimality of path planning, and the other concerns taking a detour; these are addressed in a similar algorithm to shop-preference updating. The former is dealt with by using the subjective length parameter as a weight set at each street. These weights are also updated through experience using studies such as that by Dijkstra et al. (2007), but none currently deals with the issue of time constraints. The current research devised an algorithm (similar to that of the impulse visit) in which the softmax function using a Boltzmann distribution was used to model this behavior.

The ASSA concept also considers dynamic updates within agents in order of the shortest to the longest update cycle. These are: (1) rescheduling (from 0 to several times per visit to an area, implemented in ver.1), (2) reinforcement learning of shop utility (every time after visiting an area, as implemented in ver.1), (3) updating a mental map (every time after visiting an area, the current version assumes complete knowledge), and (4) updating visiting frequency and time budget (once every few visits to an area, fixed in the current version). Another implementation possibility includes the decision-making principle (especially stance against risks and uncertainty) of an individual agent (Kaneda and Yoshida 2012).

8.3 Implementing the Osu Shopping District Case: Its Spatial Representation, Verification, and Validation

8.3.1 Framework

To evaluate the characteristics of the above designed model, ASSA ver.3 has been analyzed with consideration of the following five aspects (see Chap. 7).

1. General evaluation
   (a) Analysis of aggregated behavior (validation)
   (b) Analysis of statistics (validation)
2. Individual function evaluation
   (a) Illustration of agent's individual behavior (verification)
   (b) Analysis of similarity of visit sequence (validation)
   (c) Analysis of redundancy (validation)

This paper presents 3 as the verification examples, followed by mainly graphical representations of 1 and 2 as part of the validation results. The outcomes of other metrical analyses (such as 4 and 5) were discussed by Yoshida and Kaneda (2012).

This paper focuses on the Osu district, which is a shopping street complex district with 685 shops and facilities in eight distinct categories (cafes and restaurants, grocers, household goods, electric appliances, clothing, parks and temples, second-hand goods, and others). This district is modeled as a network with 36 street nodes (includes 4 nodes for entering and exiting the area) as illustrated in Fig. 8.2. Detailed data for both cases were obtained from a survey that was conducted during the autumn holidays in 2003 (Oiwa et al. 2005). Visitors were categorized by gender, age (three age groups), and other attributes such as place of residence and frequency of visit. In the Osu simulation case, the performance data of 2500 agents obtained

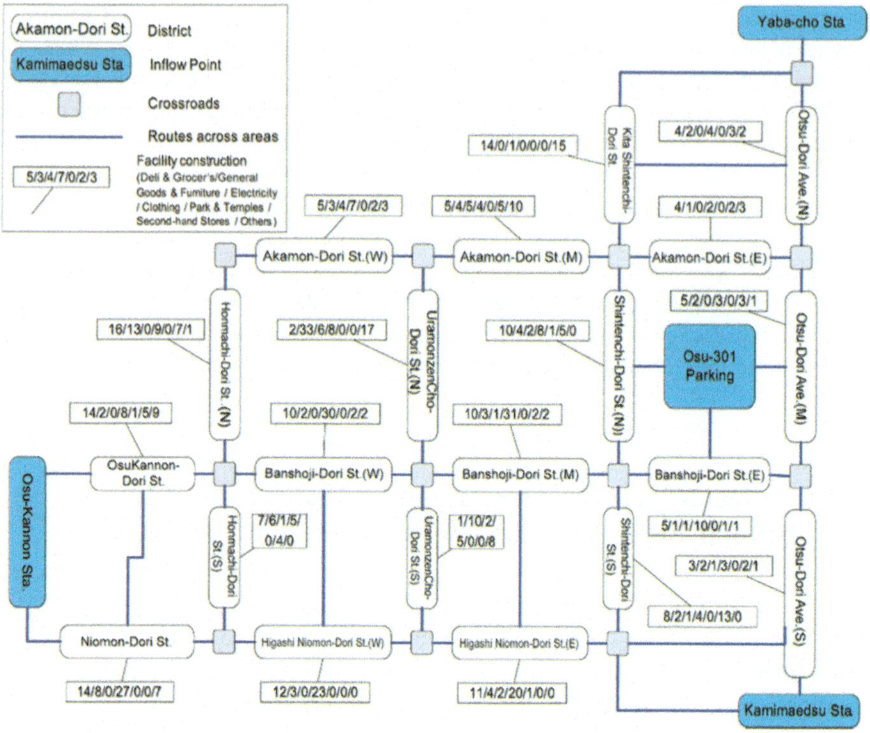

Fig. 8.2 Shop configurations in the Osu district case

over 120 days were recorded; the initial 10 days were disregarded, and the remaining data was used.

8.3.2 Verification: Examples of Agent Performance

Individual behavior results were extracted for each agent and then graphically represented on a map. The two focal points include shops visited and walking routes. For shops visited, the following four variables were considered: location, visiting order, planned or unplanned visits, and errand success or failure. Based on the observations, each of the following designed functions was verified as working: the validity of a plan created by an agent in the downtown visit decision, the implementation of the plan in the planned actions, the incidence of alternative visits, and the incidence of impulsive visits. Walking routes were mainly assessed by agent deviation from the shortest route by checking the initial planning and implementation of a planned route and the occurrence of detours from that route. The next section explains the performance of four typical agents to illustrate each elemental function described above.

8.3.2.1 Sample A

Figure 8.3 shows the tracked movements of <Agent A>. This agent represents an elderly female who lives in the <far area> and travels by train. She planned to visit a secondhand goods complex <A> (1), a clothing shop <B> (2), and a restaurant <C> (3), which she did as planned. The simulation shows that she was able to complete her errand in each shop before going home. This indicates that in this simulation, the model's planning function was executed as desired.

8.3.2.2 Sample B

Figure 8.4 shows the tracked movements of <Agent B>. This agent represents a middle-aged female who lives in the <far area> and travels by car. The agent planned to visit shops and a facility <A> (1), <B> (2), <C> (3), <D> (4), and <E> (6). She traveled to the commercial district by car. Observations show that the agent parked in parking lot Osu 301 and then visited (1). She failed to complete the errand at (1), which prompted reactional behavior. As a substitute facility, the agent visited <X> (5). This altered the plan, and facility (5) was squeezed in between (4) and (6). After this, the agent visited (2), (3), and (4). She once again failed to complete her errand at (4), and a substitute facility, <Y> (7), was selected. Subsequently, the agent visited (5) to complete the errand that was planned for (1), and

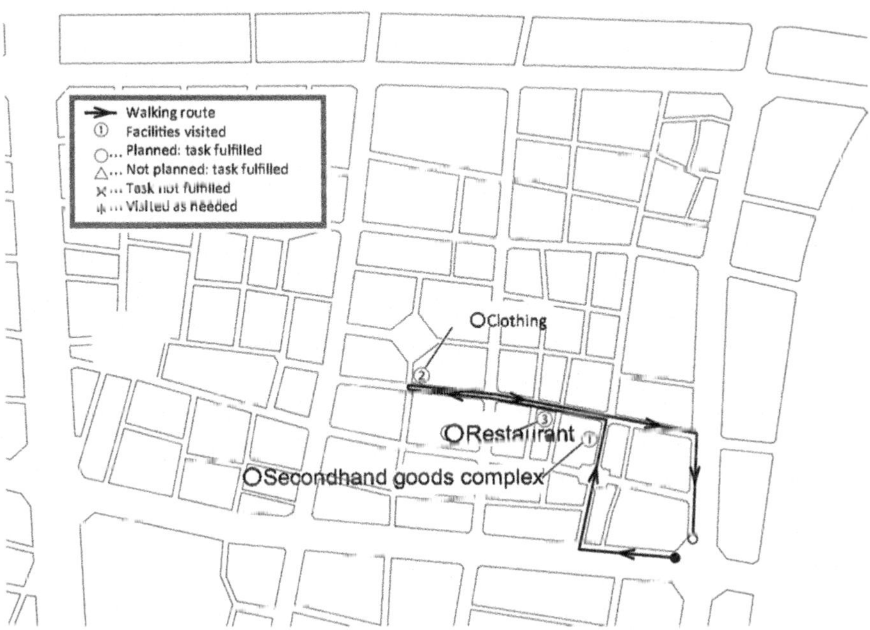

Fig. 8.3 Planned action in Agent A's performance

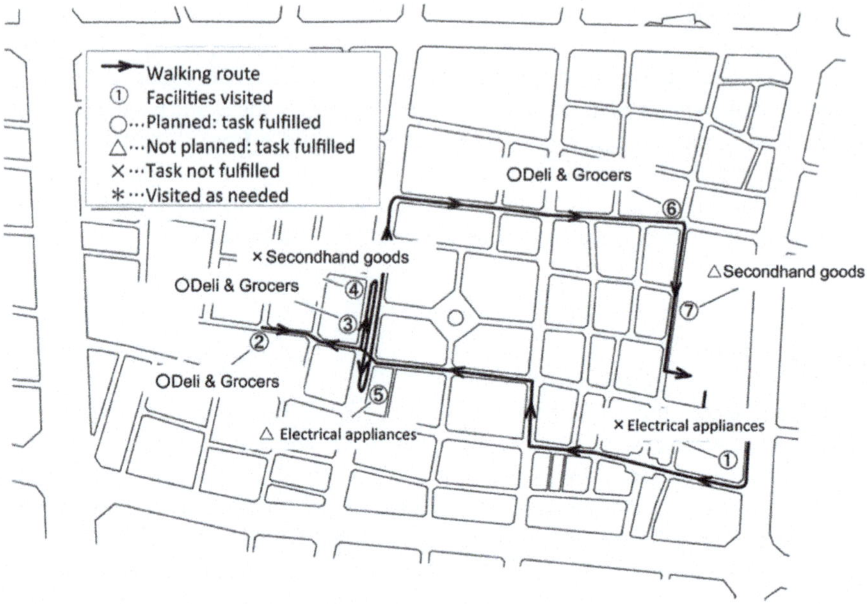

Fig. 8.4 Alternative visit and rescheduling in Agent B's performance

then visited (6) and (7). The agent then returned to parking lot Osu 301 to exit the area and she returned home. This simulation shows that the agent actively substituted her plans according to completion or failure to complete various errands at each location. The agent revised her plan as she wandered through the district.

8.3.2.3 Sample C

Figure 8.5 presents the tracked movements of <Agent C>. This agent represents an elderly male who lives in the <intermediate distance area> and travels by train to visit the district. The agent only planned to visit two shops <A> (3) and <B> (2); however, he visited two additional shops. The planned-visit facilities were <X> (1) and <Y> (4). Additionally, the sequence of visiting (2) and (3) was reversed from the initial plan because of visiting (1), which in turn altered the route. Taking a detour to visit (1) and (4) was unrelated to the two shops in the original plan. Hence, visiting these additional shops occurred as a result of improvisation behavior. The reversed sequence of visiting the facilities in the original plan can be explained when considering that the visit to (2) was time sensitive and could not be delayed. The agent satisfied the time constraint attached to (2) by delaying the visit to (3). This demonstrates the occurrence of improvisation behavior, which altered subsequent visits.

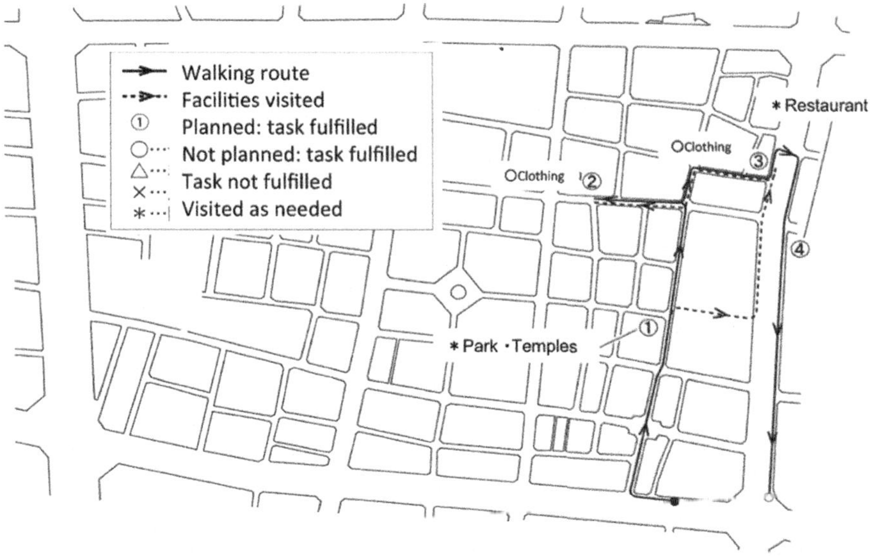

Fig. 8.5 Impulse visits in Agent C's performance

8.3.2.4 Sample D

This section examines one final agent's behavior (which was of particular note) in the Osu simulation case (Fig. 8.6). <Agent D> was a <young woman> who travels from the <intermediate distance area> by <train>. Her plan shows that she intends to visit five shops: clothing shops <A>, <B>, <C>, and <D>, and an electrical appliance shop <E>. On this agent's arrival in the district, she first visited a different clothing shop <U> (1). The agent walked along a street that is not the shortest route to shop <A>, the first planned-visit shop; this demonstrates the detour behavior function was working. Next, the agent visited a secondhand goods shop <V > (2), which again was not included in the plan. As a result of these unplanned visits, the agent changed her plan yet again and visited shop <E> (3), which she had planned to visit at the end of her trip. In the three shops, (3) and the subsequently visited <D> (4) and <C> (5), the agent successfully completed all her errands, but in <B> (6), she failed to complete her task. She chose <W> (9) as an alternative shop, but before visiting <W> (9) she visited <A> (7), which she had planned to visit earlier and then made an unplanned visit to another clothing shop <X> (8). In <W> (9) the agent completed the errand that she had failed at <B> (6) and then made unplanned visits to yet more clothing shops, <Y> (10) and <Z> (11), before finally returning home. From these observed results, it is evident that each agent action shown in Table 8.1 was working in a complex way. In addition, the impulsive-visit action of visiting several clothing shops in sequence is characteristic of the young women who frequent this commercial district; this point confirms that the model is capable of rescheduling a set of behaviors unique to an agent's attributes.

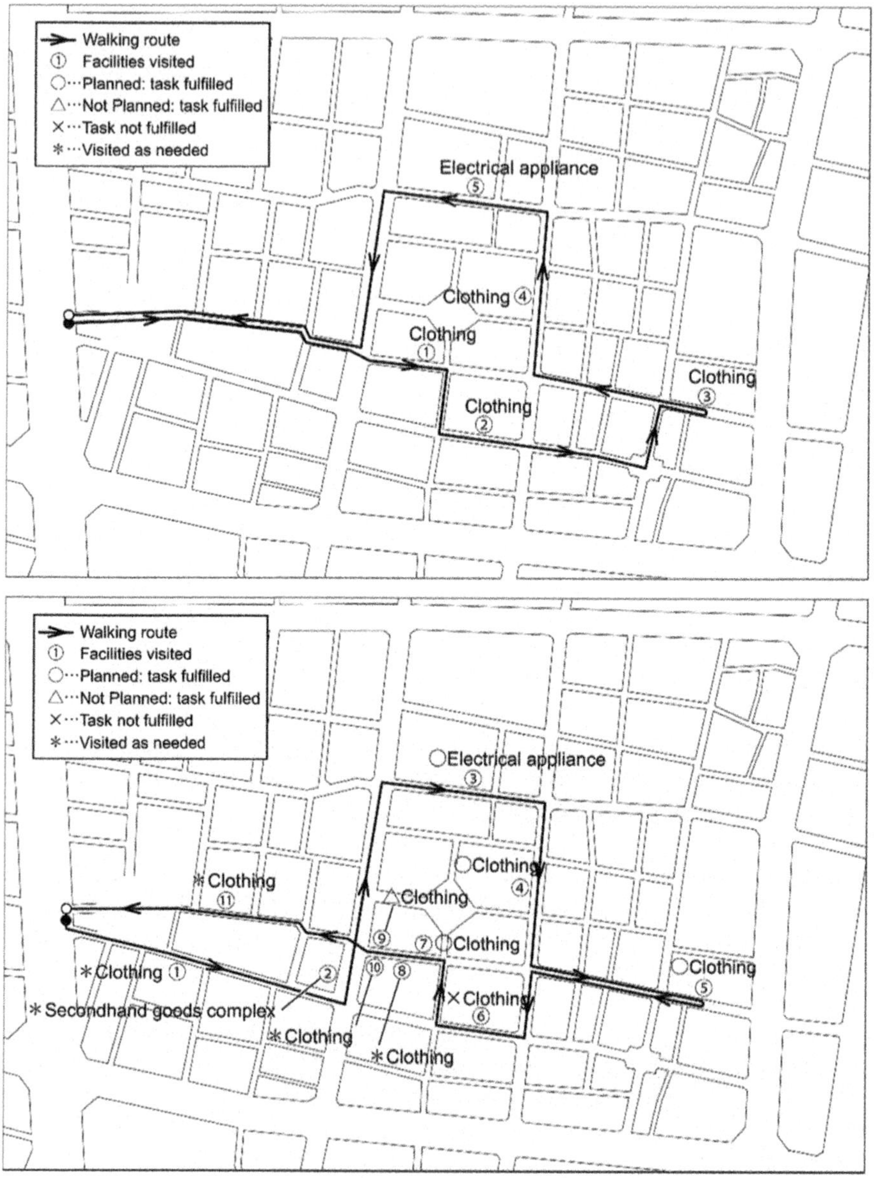

Fig. 8.6 Impulse detour action in Agent D's performance (above: plan, below: performance)

Table 8.2 Comparisons of major macro indicators

	Simulation	Survey
Duration of stay	169 min	148 min
Walk distance/walk time	18.7 min	1.15 km
Numbers of facilities visited	5.4 places	5.00 places
Numbers of planned-visit facilities	2.17 places	2.18 places

8.4 Validation: Comparison with Survey Results

Table 8.2 shows the comparison between the simulation results and the empirical data. Please note that the simulators used time distance expression, and the actual physical distance traveled was extracted from the route drawn on the survey maps, which makes comparing the two impossible. Therefore, the distance traveled is shown only as a reference. As in the previous chapter, the pedestrian's walking speed is assumed to be 1.2–1.5 m/s; the equivalent distance in the simulation is 1.35–1.68 km. For other values, it is evident that the simulation accurately predicted both the facilities visited and the facilities planned for visits. There is a discrepancy of about 10 min between the simulation and the actual data in terms of the duration of the stay in the commercial district.

The passage rate for each street and the number of facilities visited according to the simulation are shown in the lower part of Fig. 8.7. Comparing the passage rate with the empirical data displayed in the upper part of Fig. 8.7, a discrepancy is evident; the passage rate predicted by the simulation for Osu Kannon-Dori Str. ("Dori" means Street in Japanese), Niomon-Dori Str., and Honmachi-Dori Str. on the west side is below that of the actual data. Conversely, the passage rate predicted by the simulation for Kita Shintenchi-Dori Str. and Uramon Maemachi-Dori Str. is higher than that presented in the empirical data. The simulation also predicted a higher rate of visits for those facilities on the east side compared with those located on the west. This can possibly be attributed to the lack of adjustment for the agent's preference for facilities within the simulation. When choosing which amenities to visit, consideration is given to the distance between the travel mode (train station and parking facilities) and the commercial district. As such, it is understandable that the visits are more concentrated on the east side where there are more entrance points into the district. The challenges for this simulation would be to place less emphasis on distance, while improving the preference parameters for facilities and streets.

8.5 Policy Simulation Trials

To demonstrate the application of the simulation model, three scenarios are used to predict the changes in agent behavior in response to hypothetical changes made in the district and to explore the possibility of the model contributing to the revitalization of the shopping area.

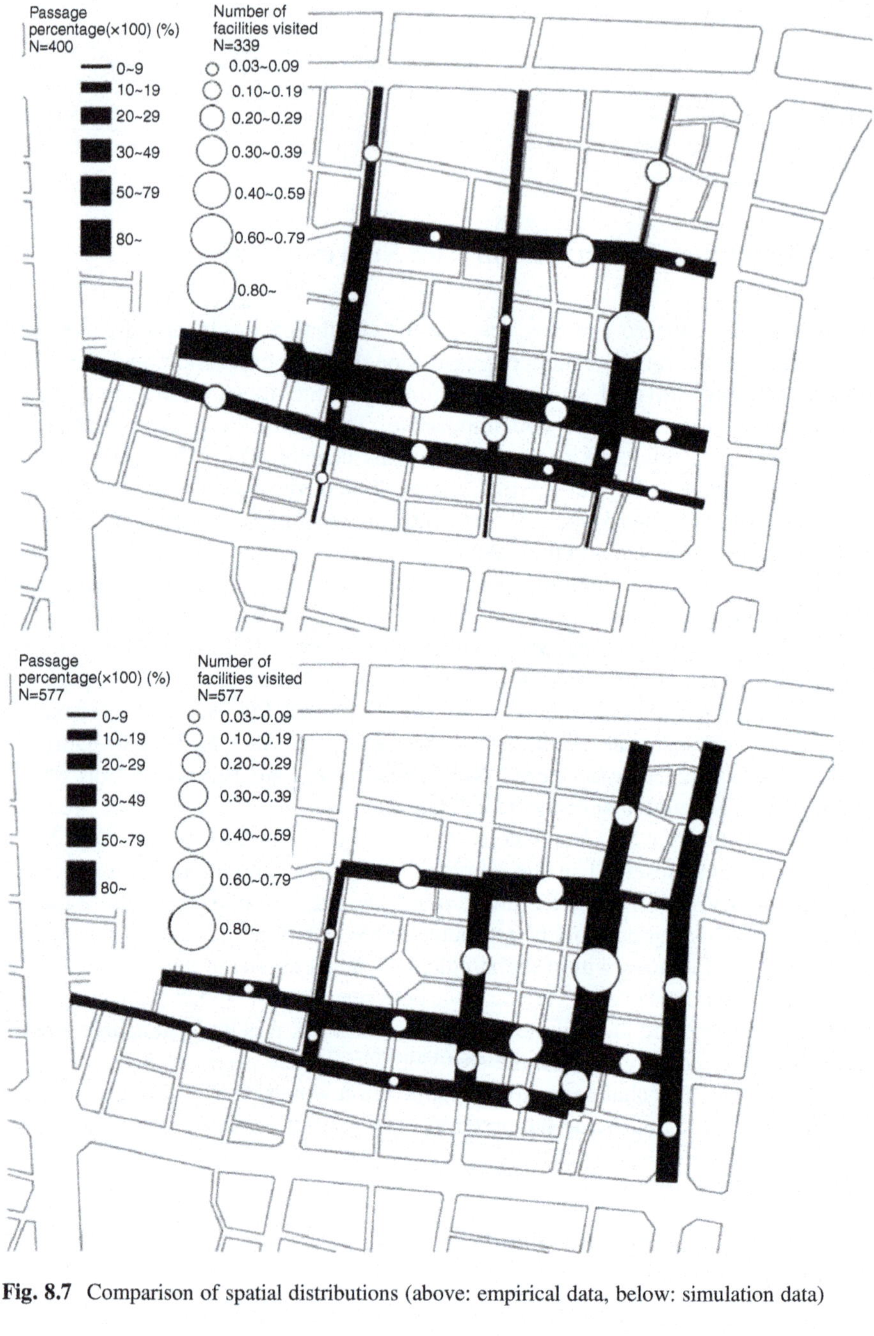

Fig. 8.7 Comparison of spatial distributions (above: empirical data, below: simulation data)

1. Scenario A: Relocation of a number of facilities
   Move a number of facilities from one street to other streets and compare the commercial district model to the base case.
2. Scenario B: Exchanging facility types
   Move all the same-type facilities from street A to street B, move all facilities of a different type from street B to street A and compare with the commercial district base case.
3. Scenario C: Strengthening a particular type of facility and increasing the background population accordingly

The number of facilities of a particular type on street A is increased compared with the base case scenario. To reflect this increase, the number of people in the background population is increased accordingly. An analysis of each case by comparing the passage rate for each street and the number of visits to facilities between each scenario and the base case is presented.

8.5.1 Scenario A: Simulation Result

Based on Scenario A, 46 shops were moved from Uramon Maemachi-Dori Str. (North), which contains 66 shops of six types (as shown in Fig. 8.2) to Honmachi-Dori Str. (north). The simulation results for passage rate and the number of facility visits is shown in the upper part of Fig. 8.8. Comparing the simulation results with the base case shown in Fig. 8.7, the number of facility visits decreased in Uramon Maemachi-Dori Str. (North) from where the facilities were moved, but the number of visits increased in Honmachi-Dori Str. (North) where the facilities were relocated. This result is rather intuitive. The passage rate was also reduced in Uramon Maemachi-Dori Str. (South), which is an area in close proximity to Uramon Maemachi-Dori Str. (north). Increased passage rates were also evident in Honmachi-Dori Str. (South), which is adjacent to Honmachi-Dori Str. (North). However, an increased passage rate was also evident in Higashi Niomon-Dori Str. (West), which is not adjacent to either of the streets in this simulation. This increased passage rate can perhaps be attributed to pedestrians walking through the area as they head to the Honmachi-Dori Str. area.

8.5.2 Scenario B: Simulation Result

Based on Scenario B, all 23 clothing shops were moved from Higashi Niomon-Dori Str. (West) (its location is shown in Fig. 8.2) to Akamon-Dori Str. (West). Higashi Niomon-Dori Str. (West) has a total of 38 shops of four different types. In the simulation, all 35 general and furniture shops were moved from Akamon-Dori Str. (West) to Higashi Niomon-Dori Str. (West). Simulation results are shown in the

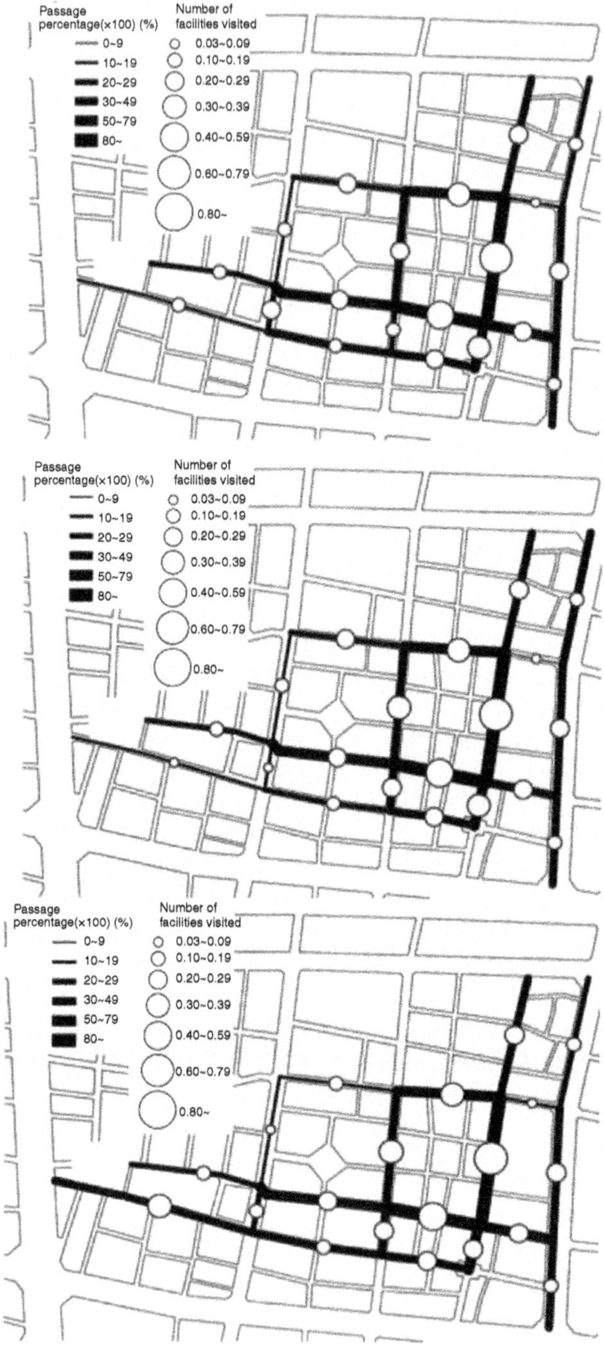

Fig. 8.8 Visitors' spatial distribution in each policy case (above: Scenario A, middle: Scenario B, below: Scenario C)

middle of Fig. 8.8. A comparison of the simulation results with the base case shown in the lower part of Fig. 8.7 reveals no significant changes to the passage rate in either Higashi Niomon-Dori Str. (West) or Akamon-Dori Str. (West). A marginal drop was observed in the number of facilities visited in Uramon Maemachi-Dori Str. (South) and Honmachi-Dori Str. (South) and in the surrounding areas of Higashi Niomon-Dori Str., but the difference was negligible. The results of the simulation indicate that provided the overall composition of the shops within a district remains constant, there would only be a marginal change to pedestrian behavior when a large number of one type of shops move out of an area.

8.5.3 Scenario C: Simulation Result

Based on Scenario C, 10 more electrical appliance shops were moved into Niomon-Dori Str. (as shown in Fig. 8.2), and the background population was increased by 250 people. The lower part of Fig. 8.8 presents the simulation results for both passage rate and the number of facility visits. Comparing the results with the base case shown in the lower part of Fig. 8.7, it is evident that the change in shop composition in Niomon-Dori Str. was accompanied by an increase in both the passage rate and the number of facilities visited. Furthermore, passage rates also increased in two areas adjacent to Niomon-Dori Str.: Honmachi-Dori Str. (South) and Higashi Niomon-Dori Str. (West). This confirms that benefits from the change extend to adjacent streets. Conversely, there was a decrease in passage rates and the number of facilities visited in Akamon-Dori Str. (West) and Honmachi-Dori Str. (North), which can be attributed to the fact that Akamon-Dori Str.'s main shops are electrical appliance shops; the increased competition from the relocated shops ultimately led to a decrease in the number of customers. Initially, it appears that the other streets were unaffected as these changes reduced visitors to the commercial district. However, the number of pedestrians and number of visits in all streets actually increased; the average number of visitors to the commercial district increased from 550 to 640 people per day due to the background population increase. The above results confirm that this scenario not only positively affected the streets to which changes were made, but a wider positive impact was felt by the Osu district as a whole. Nonetheless, the impact may be negative for those facilities located in streets that experience increased competition; therefore, this issue must be considered.

8.6 Conclusion

The ASSA model addresses downtown visitor shop-around spatial behaviors. From the perspective of a constructive approach of human-like agents, an integration of bounded rationality and feedback modifications of the agent's behaviors were explained as the key concepts of ASSA modeling. One of ASSA's main features is dynamic scheduling and includes the improvisation of downtown visitors' actions within a time budget, as opposed to existing Markov-type models.

The crucial essence of ASSA's design compared with existing approaches is a layered decomposition of the shopper agent's behaviors and their formulation as a constraint-satisfaction problem. Two topics are significant in the context of future research. One is the relaxation of the completeness of the mental map; in such cases, a utility function inside the shop-around agent would be required, although a theoretical consideration has already been devised by introducing a fundamental consideration of decision sciences (Kaneda and Yoshida 2012). Another is the parameter tuning problem. In this ASSA study, several of the parameters were tuned up in an ad-hoc manner. Artificial intelligence-aided statistical estimation methods are expected; this is currently this study's major limitation, which requires a solution.

Through these revisions, it can be concluded that ASSA would become a more useful policy and planning simulation tool.

References

Arakawa S, Kaneda T (2002) An analysis on redundancy of 'shop-around' behavior in Nagoya CBD. J Architect Plann Environ Eng (Trans AIJ) 556:227–233

Arentze T, Bogers A, Hofman F, Fujii S, Joh C, Kikuchi A, Kitamura R, Timmermans H (2001) Rule-based versus utility-maximizing models of activity-travel patterns: a comparison of empirical performance. In: Hensher D (ed) Travel behavior research, pp 569–583

Arentze T, Pelizaro C, Timmermans H (2005) Implementation of a model of dynamic activity-travel re-scheduling decisions: an agent-based micro-simulation framework, CUPUM05 paper, no. 48

Borgers A, Timmermans HA (1986) A model of pedestrian route choice and demand for retail facilities within inner-city shopping areas. Geogr Anal 18:115–128

Dijkstra J, Timmermans H, de Vries B (2007) Empirical estimation of agent shopping patterns for simulating pedestrian movement, CUPUM07 paper, no. 56

Dijkstra J, Timmermans H, de Vries B (2009) Modeling impulse and non-impulse store choice processes in a multi-agent simulation of pedestrian activity in shopping environments. In: Timmermans H (ed) Pedestrian behavior: models, data collection and applications. Emerald, London, pp 63–78

Hayes-Roth B, Hayes-Roth F (1979) A cognitive model of planning. Cognit Sci 3:275–310

Kaneda T, Yoshida T (2012) Simulating shop-around behavior. Int J Agent Technol Syst 4 (3):15–28. July-Sep, 2012

Kelly MEO (1981) A model of the demand for retail facilities, incorporating multistop, multipurpose trips. Geogr Anal 13:134–148

Kurose S, Borgers A (2001) H. A. Timmermans: Classifying pedestrian shopping behavior according to implied heuristic choice rules. Environ Plann B 28:405–418

Oiwa Y, Yamada T, Misaka T, Kaneda T (2005) A transition analysis of shopping district from the view point of visitors-shop-around behaviors. AIJ J Technol Des 22

Yoshida T (2010) A study on development of shop-around behavior simulation model of CBD visitors by using intelligent agent approach, Ph.D. dissertation, Nagoya Institute of Technology

Yoshida T, Kaneda T (2007) An architecture and development framework for pedestrians' shop-around behavior model inside commercial district by using agent-based approach. In: 10th

International Conference on Computers in Urban Planning and Urban Management, CDROM Paper-215
Yoshida T, Kaneda T (2012) The ASSA project: an intelligent agent simulation of shop-around behavior. In: The Fifth International Workshop on Agent-based Approaches in Economic and Social Complex Systems (AESCS'11), pp 148–159

Chapter 9
Modeling and Simulation of Downtown Dynamics

Toshiyuki Kaneda, Masahiro Shohmitsu, and Shuang Chang

Abstract The regeneration of "downtown" in the central business district has become an important policy, especially in developed countries. Concerning the rise and fall of downtowns, Jane Jacobs' (The death and life of great American cities, Random House, New York, 1961) concept of "city diversity conditions" is well known. From among her work, we focus on the numerous positive and negative effects of externalities that are found or can be found in diversified economic entities (e.g., shops and customers) active in downtowns.

Exploring findings concerning the rise and fall of downtowns and their sustainability, we have recently proposed Downtown Dynamics (DDy), a multi-agent social system model that focuses on the interaction between a wide variety of customer and shop agents. Thus, considering the externality of visitors who perform Multi-Purpose, Multi-Stop (MPMS) behaviors in downtowns, we propose a new errand-based visitor decision model. By abstracting ASSA, we also constructed DDy as a prototype of our downtown dynamics that handles the dynamics of introducing metropolitan areas, shop-tenant agents, etc.

This study analyzed the validity of three conditions, namely mixed primary uses, small blocks, and concentration. More work is still required for the fourth condition, aged buildings.

T. Kaneda (✉)
Graduation School of Engineering, Nagoya Institute of Technology, Gokiso cho, Showa, Nagoya, Aichi, Japan
e-mail: kaneda@nitech.ac.jp

M. Shohmitsu
Yokohama City Office, Nishi-Ward, Yokohama, Japan
e-mail: shohmitsu@city.yokohama.jp

S. Chang
Department of Computer Science, Tokyo Institute of Technology, Nagatsuda-cho, Yokohama, Japan
e-mail: chang@cs.dis.titech.ac.jp

T. Kaneda (ed.), *Downtown Dynamics*, Agent-Based Social Systems 16,
https://doi.org/10.1007/978-4-431-54901-7_9

9.1 Understanding Downtown by Agent-Based Simulation: Background and Objectives of the Research

The regeneration of "downtown" in the central business district has become an important policy, especially in developed countries. Even in the twenty-first century, when online shopping is common, downtown should function as a place where citizens visit, select their favorite products, experience novel services, or discover new things. Thus, each metropolitan area needs to have downtown. Florida (2002) has also influenced discussions on downtown policies, as required by the conditions of the twenty-first century.

Concerning the rise and fall of downtowns, Jane Jacobs' (1961) *The Death and Life of Great American Cities* is notable. In this book, she introduces "city diversity conditions" and influences debates on the planning theory and discourses that emphasize the importance of internal diversity. From among her works, we focus on the numerous positive and negative effects of externalities that are found or can be found in diversified economic entities (e.g., shops and customers) active in downtowns.

The Osu District in Nagoya City, Japan, has maintained its bustling and thriving downtown area by the ever-changing configurations of many small-scale shops (Oiwa et al. 2005; Takeuchi et al. 2011, see also Chap. 2). Based on the behavioral surveys of the conditions of the district's visitors and shops over 20 years and by adopting a bottom-up approach where diverse decision-making entities interact dynamically, the author modeled this downtown as a multi-agent system and examined its implications using simulation analyses. Although this agent-based approach deviates from the KISS (Keep It Simple and Stupid) principle, it forms the core basis of our challenging research project.

Exploring findings concerning the rise and fall of downtowns and their sustainability, we have recently proposed downtown dynamics (DDy), a multi-agent social system model that focuses on the interaction between a wide variety of customer and shop agents. Using DDy, we developed a prototype of downtown dynamics; we also conducted simulations for the sustainability of downtown commercial districts and analyzed the results. By examining the four conditions of Jacobs' concept of city diversity, we also examined the implications of our simulation analysis.

Economists have drawn on Jacobs' (1969) *The Economics of Cities*. In this book, cities are described as "the place that enables people to cooperate through learning by doing and learning from others." However, Jacobs' externality deals with communicating innovations among companies, notably manufacturing companies based in cities. The current economic texts only mention externalities on the producer side (Fujita and Thisse 2002) without any reference to theoretical models or simulation reports on city diversity conditions.

9.2 DDy Research on Jacobs' Diversity Hypotheses

In this research we set limitations on, externality in the context of commercial activity as follows has the following limitations: the secondary effects arising from any market transactions reach the third parties via means other than the transaction parties and not by way of the market. So long as an externality is a consequence of market transactions, it would be sufficient for laying the foundation at the micro-, meso-, or macro-scope level where the aggregation is possible.

The four conditions of Jacobs' city diversity, the original texts, and how our research handles each condition are classified and presented in Table 9.1.

More operationally, if we describe the research question, we would like to explore in this paper, how is it possible for the four conditions of city diversity shown by Jacobs in Downtown to be "satisfactorily" demonstrated on a model? Also, what kind of characteristics makes the model construction possible? If possible, what research programs explore key factors for this problem?

9.3 Externalities Brought About by Customers' Shop-Around Behavior in Downtown

Since externalities occur when customers and shop tenants (shops) interact in downtown, in this paper, we focus on the Multi-Purpose, Multi-Stop (MPMS) behaviors of customers and organize them into five externalities.

The first three are externalities that occur within a downtown space where visibility among customers and shops plays a central role. These externalities are as follows:

1. The "impulse visit" effect whereby a customer visiting a shop causes a visit by another shop.
2. The "town charm" effect whereby a customer's stop at multiple shops causes her or him to come to this downtown again.
3. The "bandwagon" effect whereby a customer's visit to a certain shop causes another customer to visit that shop.

The next two externalities are generated through communication and personal information devices, such as mobile terminals:

4. The "word-of-mouth" effect whereby a customer visiting a shop broadcasts this visit to another customer.
5. The "town-wide word-of-mouth" effect whereby a customer visiting a town broadcasts this visit to another customer.

These effects are shown in Table 9.2 as examples of externalities that occur to customers as micro-motive factors. Our DDy will later deal with effects (1) and (2).

Table 9.1 Texts on city diversity conditions (Jacobs 1961) and our propositions for DDy

(2) Small Blocks	– Long blocks, in their nature, thwart the potential advantages that cities offer to incubation, experimentation, and many small or special enterprises. Long blocks also thwart the principle that if city mixtures of use are to be more than a fiction on maps (p. 196) – On certain streets, any disproportionately large occupant of street frontage is visually a street disintegrator and desolator, although exactly the same kinds of uses, at small scale, do no harm and are indeed an asset (p. 247) – Frequent streets are necessary to city districts in any case, if diversity is to be generated (p. 230)	– SOS (Shop-On-the-Street) representation, maps of different sizes with same set of shop nodes are to be set initially
(3) Aged buildings (mix of old and new)	– If a city area has only new buildings, the enterprises that can exist there are automatically limited to those that can support the high costs of new construction (p. 200) – Among the most admiral and enjoyable sights to be found along the sidewalks of big cities are the ingenious adaptions of old quarters to new uses (p. 207) – In particular, they need old buildings to incubate new primary diversity (p. 208)	– For shop agent, the rent for each shop node varies
(4) Concentration (over the threshold the diversity generates)	– To be sure, the dwellings of a district need to be supplemented by other primary uses (p. 214) – In the case of enough people in enough dwellings, the diversity can be generated (p. 220) – Frequent streets are necessary to city districts in any case, if diversity is to be generated (p. 230) – The very process of increasing densities gradually but continually can result in increasing variety too, and thus can permit high ultimate densities without standardization (p. 229)	– The number of customers is to be given, that determined by a threshold value to guarantee the diversity

Table 9.2 Five externalities that customers bring by their shop-around behaviors in downtown

Externality in Town Space
(1) "Impulse Visit" Effect:
$(\exists\ c, s, s' \in T)(cVs \Rightarrow cVs')$
(2) Town Attractiveness Effect:
$(\exists\ c, s, s' \in T)(cVs \wedge cVs' \Rightarrow cVT)$
(3) "Bandwagon" Effect:
$(\exists\ c, c', s \in T)(cVs \Rightarrow c'Vs)$
Externality by communication
(4) "Mouth Words" Effect:
$(\exists\ c, c', s \in T)(cVs \wedge cIc' \Rightarrow c'Vs)$
(5) Town-size "Mouth Words" Effect:
$(\exists\ c, c', s \in T)(cVs \wedge cIc' \Rightarrow c'VT)$

Notation: c, c′: customer; s, s′: shop; *T*: Set of agents in Downtown; V, I: binary relation; cVs means "c visits s"; cIs means "c communicates with s"

9.4 Errand-Based Modeling of Customer Agents

9.4.1 Visit Decision Process in the Oligo-centric Metropolitan Area

In this section, we discuss an errand-based model similar to the assumption in ASSA (Kaneda and Yoshida (2012), Yoshida and Kaneda (2007, 2009, 2011, 2013a, b), see Chap. 7). However, our DDy assumes an oligo-centric metropolitan area in which one downtown and multiple commercial districts (such as large shopping malls in the suburbs) compete with one another. Customer agents living in various locations in this area make decisions about visiting the commercial district (Fig. 9.1).

We set some assumptions for the errands, shops, and commercial districts in downtown. First, each errand is processed in one shop, and customers cannot handle multiple errands at the same time (time occupancy). Second, shop agents' shop-type corresponds to customer agents' errand-type on a one-to-one (small shop with single function). Finally, each shop has the same size (identical). The first and second assumptions, in particular, are characterized as the so-called small shops' agglomeration district. Each of the other commercial districts competes with those in the downtown metropolitan area. These detailed assumptions are omitted and given a probabilistic achievement.

At this time, a customer's decision-making process contains three components in the following order: (1) visit time decision (when to visit), (2) visit place choice (where to visit), (3) shop-around behavior in downtown (how to act in the district?).

ASSA in Chap. 7 has already demonstrated our errands-based modeling approach: (1) "garbage can model" of visit decision time and (3) "schedule planning" of shop-around behavior in downtown.

Regarding the first component (i.e., visit decision time), the errands that arrive at a constant rate per time per errand-type will not determine a customer's daily visit every morning; however, the number of errand-types given as a threshold is

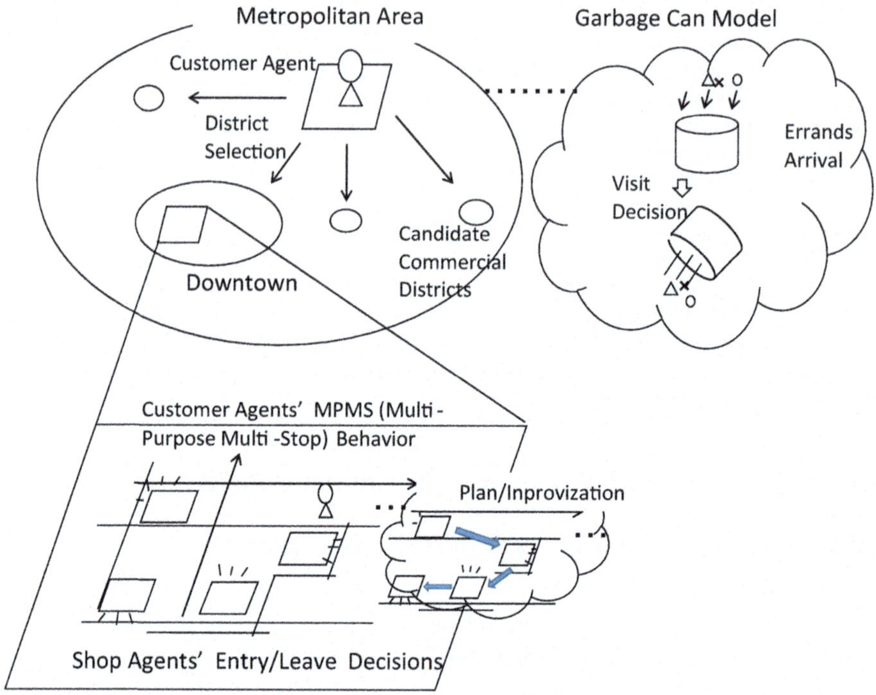

Fig. 9.1 Customer agents' decisions in DDy

provided. At that time, if the number of errand-types is the same as or above the threshold, the customer will visit the commercial district on that day to do the errands; if the number is less than the threshold, however, the customer will stay home on that day, leaving the errands undone. This model is known as the garbage can model. It should be noted that the visit interval does not become a Poisson distribution in this formulation because the threshold denotes the number of the errand-types.

9.4.2 Errand-Based Visit Place Choice Model

In the visit place choice model, the assumption of the oligo-centric metropolitan area is different from the mono-centric assumption in ASSA. Therefore, this paper introduces a new errand-based model. First, we show the basic model that replaces the commercial floor size in the logit model with two factors of commercial floor size and time distance; this is regarded as a gravity analogy with a number of errand-types of expected achievements.

When x_j is the number of errand-types for achieving the expected value of commercial district j and T_j is the time distance to there, the system's random utility term for a district visit in the basic model is as follows:

$$V_j = \alpha A_j - \beta T_j \tag{9.1}$$

(both α and β are nonnegative). The visit probability is

$$q_j = \exp(V_j) / \sum_{j' \in I} \exp(V_{j'}) \tag{9.2}$$

Considering the errand-types' number of expected achievements in commercial district j, X represents the set of errand-types that a customer has, D_j denotes the set of errand-types (shop-types) handled by this commercial district, and p_j is the average achievement number/planned number ratio of errand-types. We set x_j thus:

$$x_j = p_j \cdot \#\{X \cap D_j\} \tag{9.3}$$

Then,

$$V_j = \alpha \cdot p_j \cdot \#\{X \cap D_j\} - \beta T_j \tag{9.4}$$

However, only in downtown ($j = 0$), we replace p_j with τ:

$$x_0 = \tau \cdot \#\{X \cap D_0\} \tag{9.5}$$

If we use formula (9.4) in downtown only, then insert the walking cost of the district of the previous visit into the third term of Eq. (9.6), and consider the penalty count cost when failing to achieve the errand of the previous visit as the fourth term, we get the following formula:

$$V_0 = \alpha \cdot \tau \cdot \#\{X \cap D_i\} - \beta T_0 - \gamma \left(\widehat{L}_0 - L^*\right) - \#\text{Penalty} \tag{9.6}$$

9.4.3 *Customer Agents' Behaviors Inside Downtown*

At the time of visiting downtown, the customer agent prepares a schedule, follows a route, and improvises according to the situation. DDy follows ASSA in this sense. Table 9.3 shows the subtle differences in the mechanisms between ASSA and DDy, which is a simplified version of ASSA.

Table 9.3 Comparisons of decision mechanisms between DDy and ASSA

Stage	Decision	ASSA	DDy
Selection and decision of agent in downtown	Selection of planned shop	Choose a shop with the largest utility value (arg max utility)	Choose a shop with the largest utility value
	Plan/replan	Apply genetic algorithm considering time constraints	Always connect a pair of the nearest distance between shops (clustering algorithm)
	Errand success or failure	With probability (depend on shop-type)	With probability (depends on shop-type)
	Trigger to replan in case of errand failure	(1) When alternative shop selection (2) When impulse visit	(1) When alternative shop selection
	(a) Selection of alternative shops	With probability $1 - \varepsilon$, choose shop with the largest value regarding remained time and distance With probability ε, choose shop randomly (ε − greedy)	Step 1: Explore a planned route and choose the nearest same-type shop Step 2: Choose a shop with the largest utility value
	(b) Impulse visit of shops	[When decide] regarding the remained time, utility value of street (average value of shops' utility) [shop selection in the street] ε − greedy	[In SOS (Shop-On-the-Street) representation] Choose the one with certain probability in every shops along the route
	(c) Impulse detour	Logit model (soft max as Boltzmann, $T = 1$)	–

9.4.4 Post-updating Mechanisms Inside a Customer Agent

[Post-updating of the shop utility value]

After visiting downtown, DDy updates the shop utility value depending on the success or failure of the achievement of the errand in each of the visited shops. The visit utility function of shop i inside a customer agent is the product of reference utility value u given exogenously, errand success probability p, and correction multiplier λ_i of an endogenous variable.

Shop-Utility Updater—Stroker

$$u_i = p \cdot \lambda_i \cdot u$$

in the case of the errand success,

$$\lambda_i \leftarrow \lambda_i \cdot (1 + \delta)^{\alpha(1-p)}, \quad (\delta > 0)$$

in the case of the errand failure,

$$\lambda_i \leftarrow \lambda_i \cdot (1+\delta)^{-\alpha p}, \quad (\delta > 0)$$
$$p_i = p = 0.6, \quad \alpha = 1, \quad \delta = 0.05.$$

[Post-updating of the expected number of achieved errands]

Change the value of τ using the following formula will make it possible to asymptotically express the district utility value of downtown where its commerce composition keeps changing.

Tau (Expected Number to be Archived Errands)—Smoother

$$\tau \leftarrow r \cdot (\#\text{Bought}/\#\text{Planned Visit}) + (1-r)\cdot \tau, \quad (0 < r < 1) \quad r = 0.15.$$

9.5 Setting Up DDy: A Prototype Simulator

9.5.1 DDy as a Dynamic Simulator

DDy is a prototype of downtown dynamics that was used to conduct the simulation analysis in our research. It contains two types of agents (i.e., customer agents and shop agents) in a metropolitan area where all agents interact. In this simulator, each customer agent daily decides their visit, the district they wish to visit, and the type of behavior that best fits the errands generated.

Ten weekends/holidays are assumed as 1 month, and one simulation is conducted every 2 years (240 days). As initial conditions, six types of customer agents (a total of 50 agents) and eight types of shop agents (a total of 72 shops) are arranged on the metropolitan area map. It is important to note that eight types of shops in DDy are equivalent to eight types of errands.

9.5.2 Composition of Shop Agents

The algorithm of a shop agent (Fig. 9.2) is described below.

Each shop agent calculates the monthly profit and loss based on the accumulated number of visitors as well as other parameters (gross sales profit, fixed costs, and rent for the shop). If a shop agent has a deficit for 2 consecutive months, this agent will leave, and the shop will shut down. In the month following the 2-month deficit, the customers who pass through the street where the closed shop is located are counted, and an expected sales figure is assessed for each shop-type. When a certain shop-type with the highest figure is in the black, a shop of this type will be reopened by a new shop agent at the beginning of the following month. If no shop-type is in the black, this shop will remain shut and registered as vacant space.

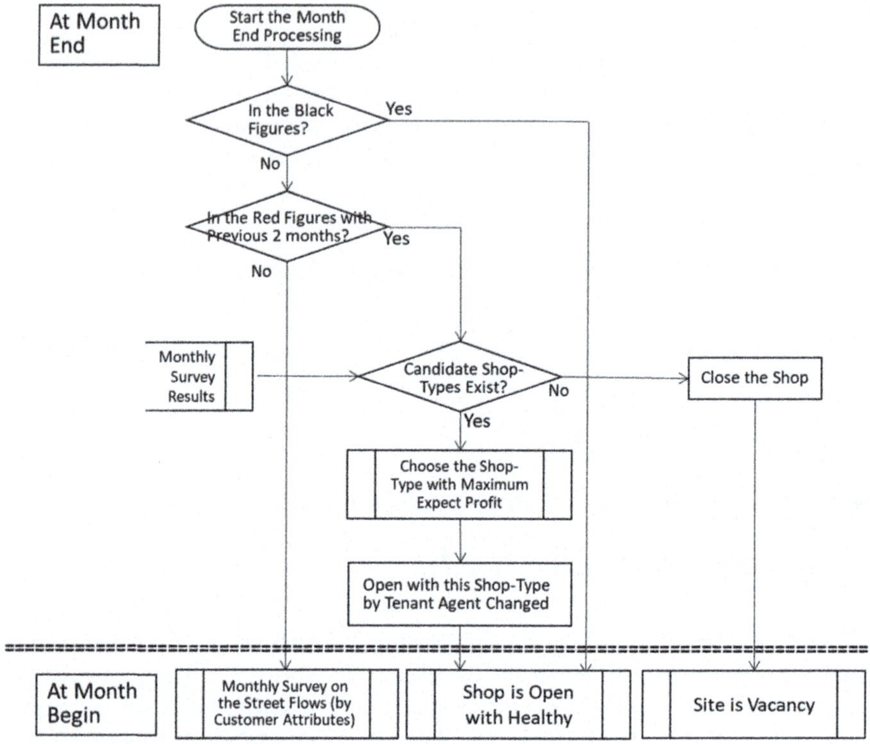

Fig. 9.2 Shop agents' Entry/Exit

A downtown map is made up of nodes marked by squares; in this map, adjacent nodes are connected by links. This map is modeled after Osu District, Nagoya City, Japan (Oiwa et al. 2005; Takeuchi et al. 2011, see also Chaps. 1 and 2). In the nodes, a customer agent shown by a circle makes a move. The Shop-On-the-Street (SOS) representation, whereby a shop is located on a node, was applied. Two downtown maps with small and large blocks containing the same 72 shop nodes were prepared (Figs. 9.3 and 9.4). To minimize the influence of any condition other than the block size, the two maps were standardized with 72 shop nodes, 16 road nodes, and 4 entrance and exit nodes. In addition, six types of customer agents and eight types of shop agents were prepared. Each type of customer agent had the same arrival ratio of the errands; thus, the tendency for visiting shop-types remained the same within each customer-type but different from the other customer types.

Fig. 9.3 A small block map

Fig. 9.4 A large block map

9.5.3 Detailed Settings of Customer Agents

At the beginning of the day, each customer agent receives new errands based on the errand arrival ratio given per customer-type per errand-type. When the number of the errand-types is equal to or higher than the threshold value, the customer agent decides on her or his visit (garbage can model). The threshold is set at 3 errand-types for all. After the agent has decided to make a visit, the parameters given to each of the four commercial districts and the downtown will be used to calculate utility value V_j for each district and select a district to visit in accordance with the logit model. When the customer agent decides to visit downtown, she or he will select a shop with the highest shop utility value in each planned errand-type; based on the shortest route principle, the customer agent will draw up a route plan. At that time, the entrance and exit nodes are congruent with the visiting means of transport of each customer agent. Upon entering a shop, the customer agent is randomly given the success or failure of the errand by assuming the errand achievement probability of $p = 0.6$. When the agent fails the errand, using the breadth-first search, she or he finds a shop of the same type from a neighboring link. Setting this result as an alternative shop, the customer agent recreates a new route plan and resumes her or his walking. Moreover, on an impulse, the agent visits a shop according to a random number with a small probability of $r = 0.03$ at each shop node. These results of the shop visits affect district visit utility value V_j and shop utility coefficient λ the next time.

9.6 Simulation Analysis of DDy

9.6.1 Simulation Case Settings

As described below, ten cases of varying conditions were prepared. The simulation was repeated 20 times for each case, followed by an analysis of the average values. A one-time simulation was conducted over 240 weekends/holidays in a population pool of 50 customer agents. The following four types of data were analyzed: (1) the number of customers per month, (2) the number of shops per month, (3) Customer City Diversity Index (CCDI, see Formula 9.7), and (4) Shop City Diversity Index (SCDI, see Formula 9.8).

$$\mathrm{CCDI} = (-1) \cdot \sum_{i \in \text{Customer Type Set}} (n_i/N) \cdot \log_2(n_i/N) \tag{9.7}$$

$$\mathrm{SCDI} = (-1) \cdot \sum_{i \in \text{Shop Type Set}} (n_i/N) \cdot \log_2(n_i/N) \tag{9.8}$$

Case 0000: A small block map was set as the basic case. The initial shop-type configuration of the 72 shops is as follows: 11, 11, 11, 8, 8, 8, 8, and 7. This arrangement means multiple (three) primary types. The agents' transport means are automobiles (30%) and public transport (70%). All values of the rents are set to 25.

Case 1000: Each number per shop-type in Case 0000 was changed using deviations in the number of shops allocated per shop-type: 36, 18, 9, 5, 3, 1, 0, and 0. This arrangement implies a single primary type. All of the other parameters are the same as in Case 0000.

Case 0100: The map was changed to conduct simulations with a large block map, as mentioned above. All of the other parameters are the same as in Case 0000.

Case 0010: The shop rent parameter of Case 0000 was changed. The rent, 15 (in 14 shops), 20 (in 14), 25 (in 16), 30 (in 14), or 35 (in 14), was randomly allocated to the 72 shop nodes to conduct simulations for a range of rents. The average value of the rents is 25, the same as the average of the rent in Case 0000. All parameters other than the rent are the same as in Case 0000.

Case 0001: In the small block map, downtown residents were settled for 40% of the customer agents to conduct simulations for a case to promote living inside downtown. The means of transport were set as follows: automobile 18%, public transport 42%, and walking (of residents) 40%. The downtown residents were randomly positioned on the shop nodes. The time distance to downtown was considerably reduced from 45 to 5 min.

Case 0101: For the large block map, similar to Case 0001, simulations were conducted for a case to promote living inside downtown.

9.6.2 Analysis of Condition 1: On the Need for Mixed Primary Uses

Case 0000 and Case 1000, where the number of shops was changed for each shop-type, were compared. The results are shown in Fig. 9.5 and explained below.

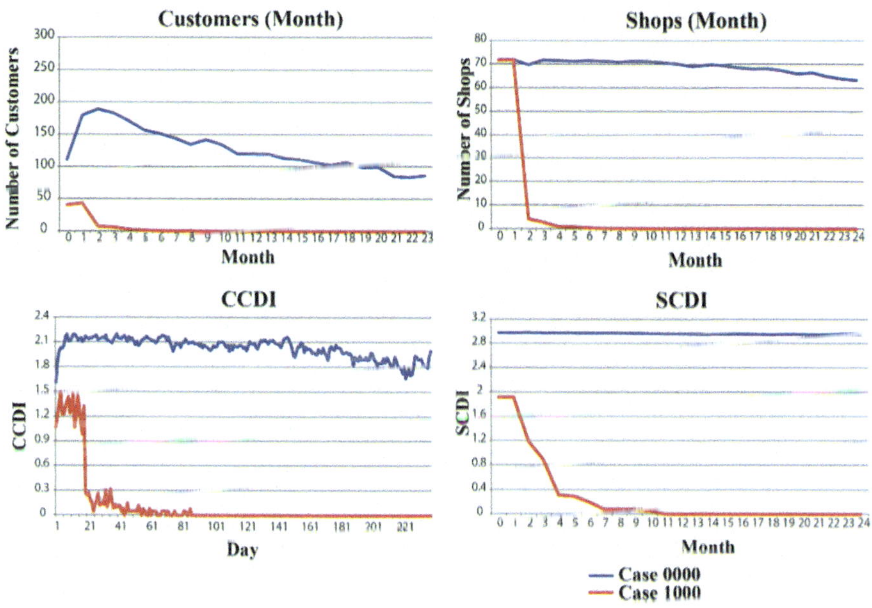

Fig. 9.5 Simulation results of condition 1

In Case 0001, where a single primary type was given as the number of shops in each shop-type, both the numbers of customers and shops started declining at an early stage. This is because an unbalanced shop-type configuration triggered a loss of diversity from the early stage, which is also shown by CCDI and SCDI.

9.6.3 Analysis of Condition 2: On the Need for Small Blocks

Case 0000 and Case 0100, where the block sizes differ, were compared. The results are shown in Fig. 9.6 and explained below.

When the numbers of customers were examined in Case 0000 with small blocks and Case 0100 with large blocks, the numbers tended to decrease; however, Case 0000 always showed a higher number. There was a sudden decrease in the number of shops in Case 0100; in contrast, the number hardly decreased in the small blocks where a great difference was seen at the end of the simulation. This tendency was also noticeable in CCDI and SCDI. This means that specific types of customers and shops decreased in Case 0100, introducing bias. In contrast, they decreased evenly in Case 0000 with no bias. Hence, it can be argued that Case 0000 maintained more diversity.

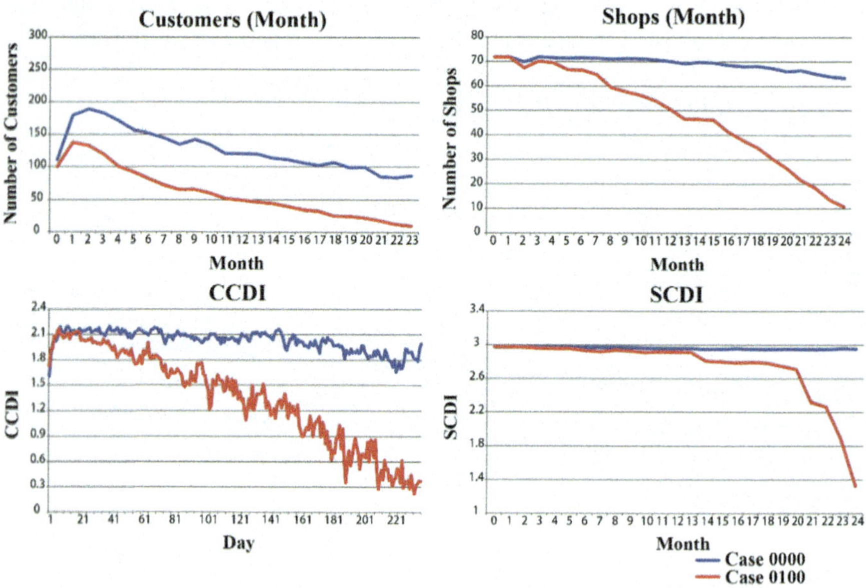

Fig. 9.6 Simulation results of condition 2

9.6.4 Analysis of Condition 3: On the Need for Aged Buildings

Case 0000 was compared with Case 0010, where different rents were set. The results are shown in Fig. 9.7 and explained below.

Virtually no difference was found in the number of shops; however, in SCDI, the values of Case 0010 with different rents dramatically declined, introducing bias into the shop-type configuration. Moreover, the values of Case 0010 were lower for both CCDI and customer numbers. These results are different from the conditions suggested by Jane Jacobs, which could emanate from an inconsistency in the combination of the break-even point set for each shop-type and the given rent arising from the random setting of the rents to the shop nodes. This is an ongoing analysis that requires constant monitoring. Nonetheless, the simulation results strongly support Jacobs' assertion that if diversity is lost, downtown will fall into a decline.

9.6.5 Analysis of Condition 4: On the Need for Concentration

In the small block map, Case 0000 and Case 0001 with their different residential densities were compared. The results are shown in Fig. 9.8 and explained below. The comparative results are also shown in Fig. 9.9.

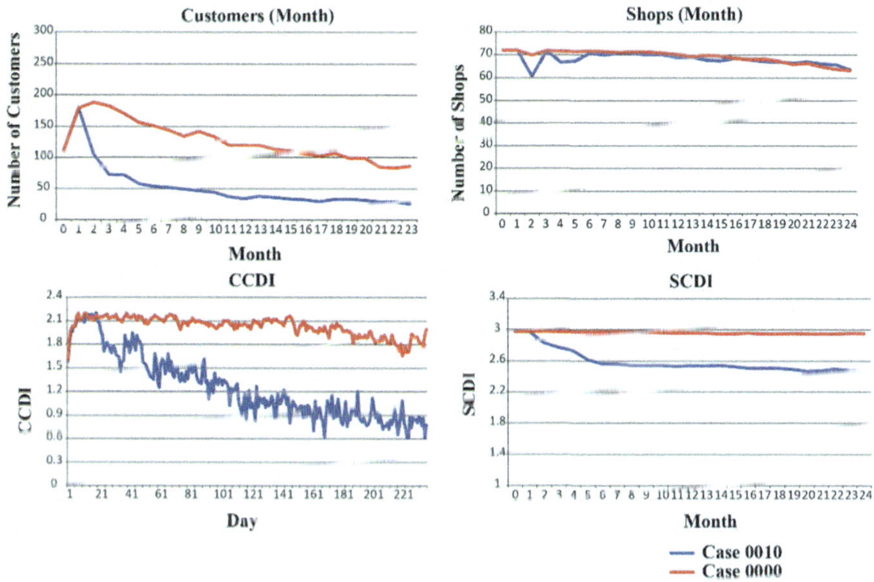

Fig. 9.7 Simulation results of condition 3

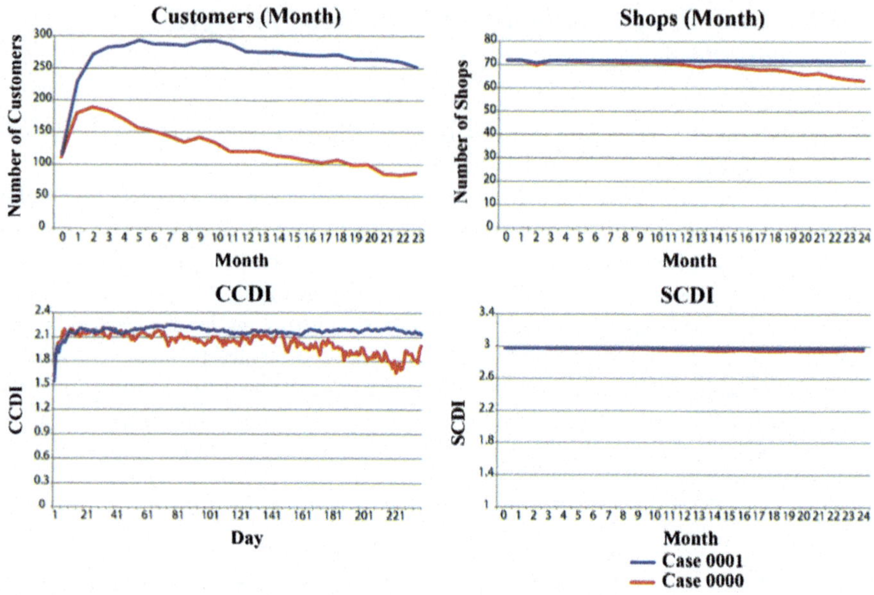

Fig. 9.8 Simulation results of condition 4 (in small block cases)

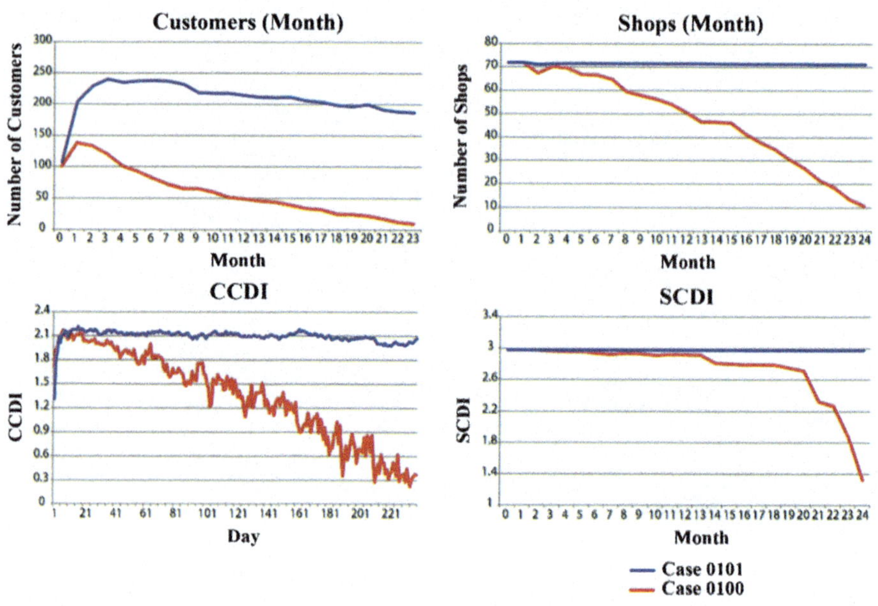

Fig. 9.9 Simulation results of condition 4 (in large block cases)

Case 0101, where downtown residents were set, had values far above Case 0100 in both customer and shop numbers, preventing downtown decline for covering the disadvantages of large blocks, as indicated in Condition 2. The values of Case 0101 were also considerably higher in CCDI and SCDI. Therefore, it can be argued that a higher density and concentration of downtown residents can more readily maintain diversity. It was also found that the downward trend in the number of customers was slowed in the case of large block maps.

9.7 Conclusion: Implications of DDy Simulation Analysis

In this chapter, we focused on Jacobs' city diversity hypotheses to examine the sustainability conditions of downtown. Accordingly, based on the externality of visitors who exhibit MPMS behaviors in downtown, we proposed a new errand-based visitor decision model. By abstracting ASSA, we also constructed DDy as a prototype of our downtown dynamics that handles the dynamics of introducing metropolitan areas, shop-tenant agents, etc.

The findings of our simulation analyses are discussed below.

For the first analysis, we conducted simulations and compared between one case with multiple primary uses and another case with a single primary use. The results confirm that the case with multiple primary uses preserves the diversity and sustainability of downtown longer, supporting the need for mixed primary uses.

For the second simulation analysis, we compared between a small and a large block case. The results confirm that the small block case is sustainable and maintains more diversity, supporting the need for a small block—Jacobs' second condition.

For the third analysis, simulations were compared between one case where the same rent was set and another case where the rent was set differently. As a result, the case where the rent was the same became dominant. In this simulation analysis, we did not receive sufficient support for the third condition.

For the fourth analysis, we conducted simulations in two case settings on the presence or absence of residents in both of the small and large block towns. No major difference was noticed in the small block town. Nonetheless, the results confirm that the large block town helps in maintaining the diversity and sustainability of downtown by settling downtown residents. These findings provided support for the fourth condition (i.e., dense).

This study managed to analyze the validity of the first, second, and fourth conditions of the four city diversity conditions proposed by Jane Jacobs. It needs to be noted that there remains a problem in the simulation model design in that it presents the new and old buildings of the third condition as differences in the shop rent.

DDy's customer-based decision model lacks grounding because of limited empirical data. However, we wish to emphasize that the proposal of a new model through exploratory research is in itself a consistent means of expression that encourages thinking to obtain new insights. We also would like to argue that this model is susceptible to falsifiability in that new grounding may cause the components of the model to be modified. Hence, there are great prospects for further research.

The general consensus is that a DDy model that deals with Jacobs' diversity conditions can be constructed through the agent-based approach. Those who live in a city feel something like a "symptom" for the ups and downs of their downtown, although this is a very personal feeling. However, DDy, which is a bottom-up approach, can be used as a model to reproduce such simulated situations for fruitful discussions. On that note, the authors deem it impossible to prevent the entry of this prototype simulator into our artificial society.

Acknowledgments A part of the research received Grants-in-Aid for Scientific Research, 26282083 and 25240048, from the Japan Society for the Promotion of Science (JSPS).

References

Florida R (2002) The rise of the creative class. Basic Books, New York

Fujita M, Thisse JF (2002) Economics of agglomeration. Cambridge University Press, Cambridge

Jacobs J (1961) The death and life of great American cities. Random House, New York

Jacobs J (1969) The economy of cities. Random House, New York

Kaneda T, Yoshida T (2012) Simulating shop-around behavior. Int J Agent Technol Syst 4 (3):15–28

Oiwa Y, Yamada T, Misaka T, Kaneda T (2005) A transition analysis of Shop-Ping District from the view point of visitors' shop-around Behaviors—a case of Ohsu District, Nagoya. AIJ J Technol Des (in Japanese)

Takeuchi M, Yoshida T, Kaneda T (2011) A transition analysis of shopping street complex district from the viewpoint of shop-around Behaviors—a survey of Osu District, Nagoya in 2008. J Architect Plann Trans AIJ (in Japanese)

Yoshida T, Kaneda T (2007) An architecture and development framework for pedestrians' shop-around behavior model inside Commercial District by using agent-based approach. In: 14th International conference on computers in urban planning and urban management

Yoshida T, Kaneda T (2009) A simulation analysis of shop-around behavior in a Commercial District as an intelligent agent approach—a case study of Osu district of Nagoya City. In: Agent-based approaches in economic and social complex system V, post-proceedings of the AESCS international workshop 2007

Yoshida T, Kaneda T (2011) ASSA: an intelligent agent-simulation for shop-around behavior, focusing on its performance evaluation. In: 12th international conference on computers in urban planning and urban management

Yoshida T, Kaneda T (2013a) ASSA: project: an intelligent agent-simulation for shop-around behavior evaluation. In: Agent-based approaches in economic and social complex systems VII, post-proceedings of the AESCS international workshop 2012

Yoshida T, Kaneda T (2013b) Policy simulation trials of intelligent shop-around agent model. In: 13th international conference on computers in urban planning and urban management

Part IV
Emergence of Vision-Driven Agents

Chapter 10
The Potential of Vision-Driven Agent Simulation: The VD-Walker

Kasumi Susaki and Toshiyuki Kaneda

Abstract Recently, the decline of downtown areas has become a major issue for Japan, and no effective strategies to solve the problem have been found. The revitalization of downtown areas is one of the targets of urban planning, but there is no established method for measuring crowds. Space Syntax is one analysis theory for finding strategies that make cities livable. This theory has been implemented as a tool that works on Depthmap software. Depthmap includes several tools, one of which is called EVA, the pedestrian agent simulator. EVA was designed and developed by Turner and Penn. Urban districts should be designed from the perspective of pedestrians. Thus, the VD-Walker, a vision-driven agent simulator that redesigned EVA's model and implemented it on UNITY, a three-dimensional (3D) game engine, was developed. In this paper, authors first explain Gibson's "affordance" and Hillier's "natural movement" concepts, which EVA incorporates. Second, authors describe the algorithm of the current version of the VD-Walker and its basic functions, and several simulation examples are shown. To conclude, Gibson's suggestive affordance theory will be used to create a new model in UNITY's 3D environment, and the potential of UNITY as a simulation platform and future research on the VD-Walker will be discussed.

10.1 Introduction

In Japan, it has become increasingly popular for residents to own their own cars, and big suburban-style shopping centers are becoming common. Because of this situation, historical downtowns are in decline. This tendency has been

K. Susaki (✉)
Ryukoku University, Kyoto, Japan
e-mail: susaki@world.ryukoku.ac.jp

T. Kaneda
Graduate School of Engineering, Nagoya Institute of Technology, Gokiso, Showa, Nagoya, Japan
e-mail: kaneda@nitech.ac.jp

T. Kaneda (ed.), *Downtown Dynamics*, Agent-Based Social Systems 16,
https://doi.org/10.1007/978-4-431-54901-7_10

recognized as a major issue, but no effective strategy to solve the problem has yet been found.

Because Japanese society is facing several critical issues, such as the birth of fewer children, more elderly people, and a decreasing overall population, the Japanese government has proposed a new policy. This policy, called the Location Normalization Plan, included in the Act on Measures Concerning Urban Reconstruction, is currently being implemented. It highlights the importance of walkable urban districts where residents can do their shopping, etc.

10.2 Vision-Driven Pedestrian Agent Simulation

10.2.1 Designing Districts Based on the Analysis of Pedestrian Behavior

The design model in which vehicles have top priority should be changed to one in which pedestrians have top priority. To preserve or revitalize downtown areas, the modeling of pedestrian behavior is effective. Traveling behavior surveys and similar research methods have been conducted, but the results have been recognized only as case studies.

In the field of architecture, there have been many surveys of resident behavior in front of stations or in squares, and the results of these surveys and the research results of psychological or behavior analysis have been used as demonstration data or theory.

Kaneda et al. (2020) reported a case study at Kanayama District in Nagoya City. This case study conducted both an encounter survey and analysis by VGA, one of the tools of Depthmap and EVA. The results of EVA showed a higher correlation than VGA with the results of the encounter survey.

These results show that when designing districts in urban planning (meso-level), the perspective from pedestrian behavior (micro-level) is important.

10.2.2 Vision-Driven Pedestrian Models

There are only a few models that simulate pedestrians' daily behavior in their districts and many models that simulate optimal path selection or evacuation behavior in an emergency.

An agent simulation model introduced by Penn and Turner (2002) that simulates pedestrian behavior was designed to use more detailed (i.e., micro-level) analysis than VGA. This pedestrian agent simulation is the basis of EVA (Turner and Penn 2002). Space Syntax is based on the line of sight, and EVA was designed on Space Syntax. EVA is a model that unites Gibson's research (Gibson 1978) about visual perception, especially for "natural vision" and "affordance," with Hillier's "natural movement" (Hillier et al. 1993).

In the EVA model, human visual perception directly affects pedestrian behavior. This kind of model was named the vision-driven model by Lee et al. (2013). Natural movement was explained as a basic behavior that is useful in understanding human behavior based on visual perception. The base level behavior, such as visual perception, should have been met before upper-level behavior such as decision-making would affect human behavior, so it was very important to understand vision-driven human behavior. Natural movement is a phenomenon that affects human behavior in a built environment, and simulation has become available to validate it.

Kaneda et al. (2019) defined a vision-driven agent as an agent that moves according to visual perception. The method and timing of directional decision are varied and important factors in a vision-driven agent system. A vision-driven agent model has a black box that excludes visual perception and includes other information or any kind of recognition that the agent uses to choose its walking path. In this sense, the vision-driven agent is a basic model.

10.2.3 EVA

The agent model EVA proposed by Turner and Penn (2002) is a vision-driven agent model that united Gibson's natural vision and affordance (Gibson 1978) with Hillier's natural movement (Hillier et al. 1993). This model explained that "the human moves in a direction that provides him or her the potential for possible further movement" (Turner and Penn 2002). It also referred to Gibson's work, "When no constraints are put on the visual system, we look around, walk up to something interesting and move around it so as to see it from all sides, and go from one vista to another. That is natural vision ..." (Gibson 1978: 1) and commented that natural vision is a combination of factors that affect human behavior. Natural vision has been explained by Turner and Penn (2002; 474) as a similar behavior to "we look around" and "we go from one vista to another." It is similar to natural movement. In this sequence, natural movement was explained by referring to the phrase "an agent does not even require the ability to recognize 'object' as distinct from 'environment'; the agent merely has to recognize that there is environment which may be explored in order to move (though not necessarily to navigate)" (Turner and Penn 2002: 474). This natural movement was then described by Hillier's (Hillier et al. 1993) expression that most movements of pedestrians follow the length of the line of sight, and pedestrians connect the line of sight to the next line of sight as they walk.

The agents of EVA move on a plate overlapped with grids that have position and information about the available directions for the agent on the grid. In EVA, 360° is divided into 32 bins, each with a fixed, equal angle of 11.25°. The EVA agent's standard field of view is 15 bins, i.e., 170°. This angle was chosen because it allows pedestrians to choose to go straight, turn left, or turn right at the crossing (Penn and Turner 2002). When a pedestrian looks around from left to right while walking, the field of view is 170° (AIJ 1980).

The direction in which to go is chosen by probability proportional to the area of the bins. Once an agent chooses a direction, it moves the designated steps that were decided by probability, using a Poisson distribution under the length of the line of sight. Then, the agent chooses its next direction. When the agent has no direction in which to go at least one step in the field of view, the agent randomly chooses to turn 90°left or right, and when the agent can move at least one step in that direction, the agent chooses the direction to go and moves the designated steps. The number of steps is chosen in the same way as usual movement. When an agent can move to neither side, an arbitrary angle from 0° to 360° is chosen. The speed of the EVA agent is 1.5 m/s, in accordance with human walking speed. Turner and Penn (2002) reported a good correlation between the encounter survey and simulation results in the case study of Tate Gallery and a district in London.

10.2.4 Weighted Origin Destination Model

An open problem of EVA agents is that in a square or similar place, direction selection does not work well, and the agent continues to move inside the square for too long. To solve this problem, a weighted origin destination model was proposed by Ferguson et al. (2016). In this model, each agent first deletes the bins that digress from the destination and chooses a direction from the remaining bins. The timing and method of direction choice are the same as EVA's.

An agent in the weighted origin destination model has redundancy with EVA's model but moves toward the destination.

10.2.5 Three-Dimensional Vision-Driven Agents

EVA is a vision-driven pedestrian model on a plate, probably because of the constraints of computer power at that time. But Gibson's research on visible perspective (Gibson 1978) can be applied to human daily life and is applicable for space that includes height.

The natural vision that Turner and Penn referred to explained that people moved from vista to vista and showed that Gibson's research target was a space that included height.

Kaneda et al. (2019) proposed VD18s, a 3D vision-driven pedestrian agent simulation model implemented on the 3D game engine UNITY and reported the simulation results. The VD18s used not only line of sight but also a checking mechanism of obstacles to walk around at a low level. The VD18s introduced comparable results to human behavior and psychological analysis.

10.3 The VD-Walker

10.3.1 Implementation on the 3D Game Engine UNITY

UNITY has enough expressive 3D power to be used as a space model for 3D computer-aided design systems for pedestrian agent simulation. Though other simulation platforms such as Net Logo also realize 3D space expression, UNITY has more potential for 3D space expression and can create 3D space models easily.

For these reasons, the VD-Walker was redesigned on UNITY. The basic algorithm of the VD-Walker is nearly the same as EVA's. EVA's agents move on a plane covered with grids, and agents get position and other information by grid. Agents of the VD-Walker move on 3D space that has *x*-, *y*-, and z-axes. The EVA algorithm was modified, and polar coordinates were introduced to each agent. The algorithm of the VD-Walker, like EVA's, set the spacing between grids to infinitesimal.

The simulation on UNITY can be seen in Fig. 10.1. Red dots show agents and white dots show the trajectory of agents.

In the VD-Walker, the viewing angle is called the field of view, as in EVA. The angle can be set as a parameter before the simulation starts. The standard field of view is 170°.

In the left side of Fig. 10.2, the red mark shows an agent, and its field of view is in front (upper side in the figure) of the yellow lines. The agent chooses a direction and moves the designated steps. The right side of Fig. 10.2 shows the agent repeating direction selection and moving the designated steps.

Figure 10.3 shows the bin and line of sight. The bin of the VD-Walker divides the field of view into the same angle (black lines), and there are seven bins in this figure. The red lines are marked in the center of each bin and show the directions that this agent can choose to move. These red lines are called the line of sight. The black circle shows an agent with a field of view of seven bins, and the agent can choose one direction from among these seven.

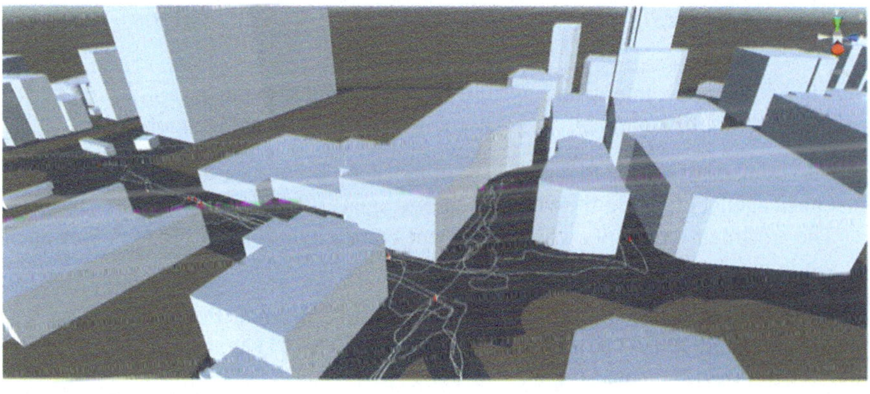

Fig. 10.1 Simulation snapshot of the VD-Walker on UNITY

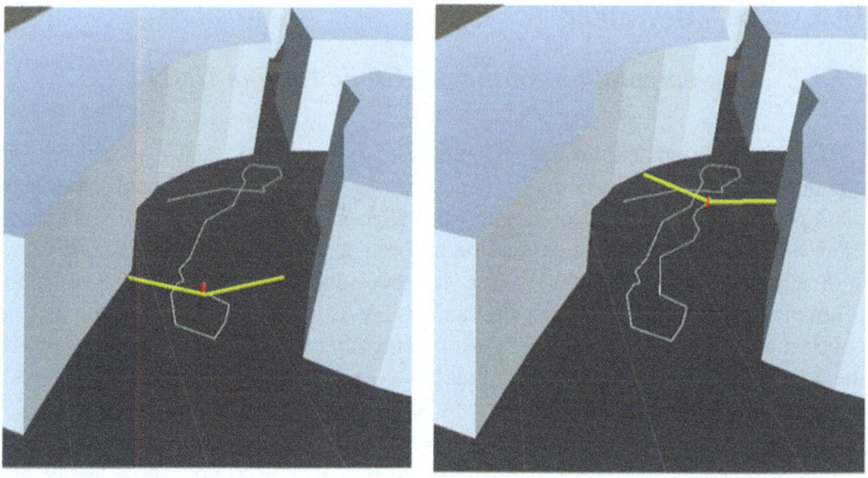

Fig. 10.2 Field of view

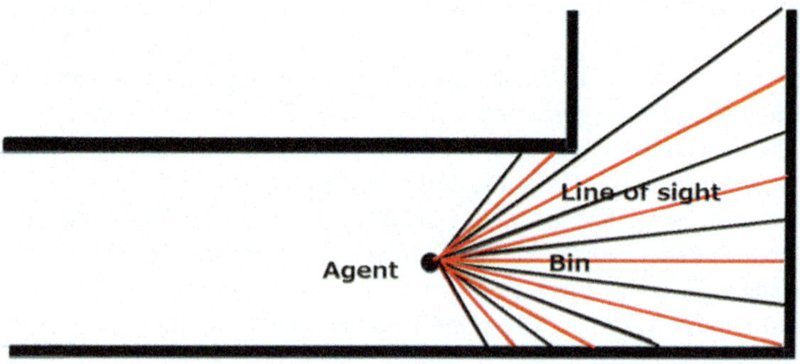

Fig. 10.3 Bins and lines of sight

10.3.2 The Design Issue of Interaction Between Agents

When designing an interaction system between agents in the vision-driven pedestrian agent simulation model, the following three items must be addressed:

1. The number of agents that can exist at the same point.
2. Other agents blocking the view when one agent moves.
3. The treatment of agents when one agent comes into contact with another agent while moving.

The comparison of the three models (EVA (Turner and Penn 2002), the wall vector field model (Lee et al. 2013), and the VD-Walker) regarding the abovementioned items is presented in Table 10.1.

Table 10.1 Comparison of agents' interaction

	EVA	Wall vector field model	The VD-Walker
Number of agents that can exist at the same point	Only one	Only one	Only one
Other agents block the view when the agent moves	No. Agents are transparent to other agents	Yes. Agents are treated as visual obstacles to their agents	Either is possible. The default is "No," but can be changed to "Yes" using parameters
When an agent contacts another agent while moving	An agent can walk through other agents	Agents behave to avoid physical interaction	Agents crash sometimes

In the EVA model, agents have the characteristic that other agents do not have substance, and thus agents can walk through other agents while moving. In the current version, agents in the VD-Walker cannot walk through other agents. Other agents of the VD-Walker block the view when the agent moves, but this can be changed by changing the parameter settings. When this setting is changed so that other agents do not block the view when an agent moves, but the agent cannot walk through other agents, crashes can occur. The agents of the wall vector field model are treated as visual obstacles to other agents.

The VD-Walker includes many parameters, including UNITY's, that define agent behavior. For example, field of view, number of bins, speed, length of step, the average of Poisson distribution used to decide number of steps, etc. can be set for each agent.

10.3.3 *Algorithm of the VD-Walker*

The algorithm of the VD-Walker is shown in Fig. 10.4. Details are shown below.

Procedure 1

Set a target point.

Probability of direction of destination:

$$p_i = \frac{l_i^2}{\sum_1^n l_i^3}$$

Probability of travel distance to destination:

$$Q_i = \frac{i^2}{\sum_1^n i^2} \leq 1_i$$

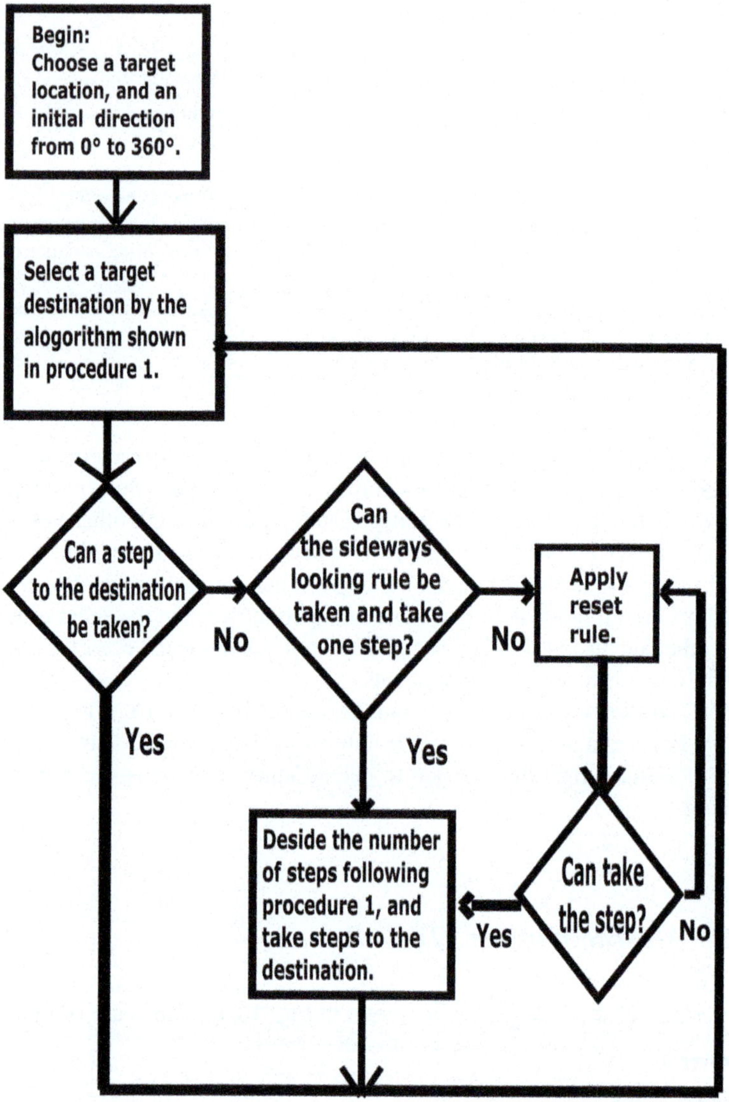

Fig. 10.4 Algorithm of the VD-Walker

In these mathematical expressions, n is the number of bins and L is the length of the line of sight. The movement distance is determined by probability using the Poisson distribution from one step to the maximum number of steps I_{max}.

Procedure 2
At each step, it is determined whether the target point and the number of straight movements have been reached. If so, then the process proceeds to procedure 1.If not, then procedure 2 is performed.

Procedure 3 (Irregular Case)
In the case of a wall collision, apply the sideways-looking rule (select 90° right or left direction). If the agent still cannot move, apply the reset rule (choose an angle randomly from 0° to 360° as a new direction) and execute procedure 1.

The VD-Walker agent decides the distance to move by the number of steps. When the selected direction does not have enough length to take one step, procedure 3 is immediately executed.

The algorithm of the VD-Walker uses the sideways-looking rule. Using this rule, agents move toward the selected direction not only one step but several steps according to Q_i and probability using the Poisson distribution in procedure 1. This procedure is different from EVA's. According to EVA's algorithm, in the case of a wall collision, the agent checks whether it can take one step 90° to its right or left. When the agent can take a step 90° to its right or left, it does so. After the agent takes one step, it decides whether or not to change direction based on a Poisson distribution with mean n.

10.3.4 Navigational Movement Model of the VD-Walker

The weighted origin destination model of EVA was introduced to the VD-Walker. The algorithm was modified to adjust to the VD-Walker. The algorithm of the navigational movement model of the VD-Walker is shown in Fig. 10.5. Details are shown below.

Procedure W1
Redefine the field of view of the agent following to the target.

Mark the lines where the line from the agent to the target rotates 90° to the right and left. The overlapping area of the field of view and the 180° area is the area for the subset of bins.

Procedure W2
When there is no subset, rotate the agent with a random angle from 0° to 360° and go to Procedure W1.

Procedure 1
Set a target point.

Probability of direction of destination:

$$p_i = \frac{l_i^2}{\sum_1^n l_i^3}$$

Probability of travel distance to destination:

$$Q_i = \frac{i^2}{\sum_1^n i^2} \leq 1_i$$

In these mathematical expressions, n is the number of bins and L is the length of the line of sight. The movement distance is determined by probability using the Poisson distribution from one step to the maximum number of steps $I_{\max}$.

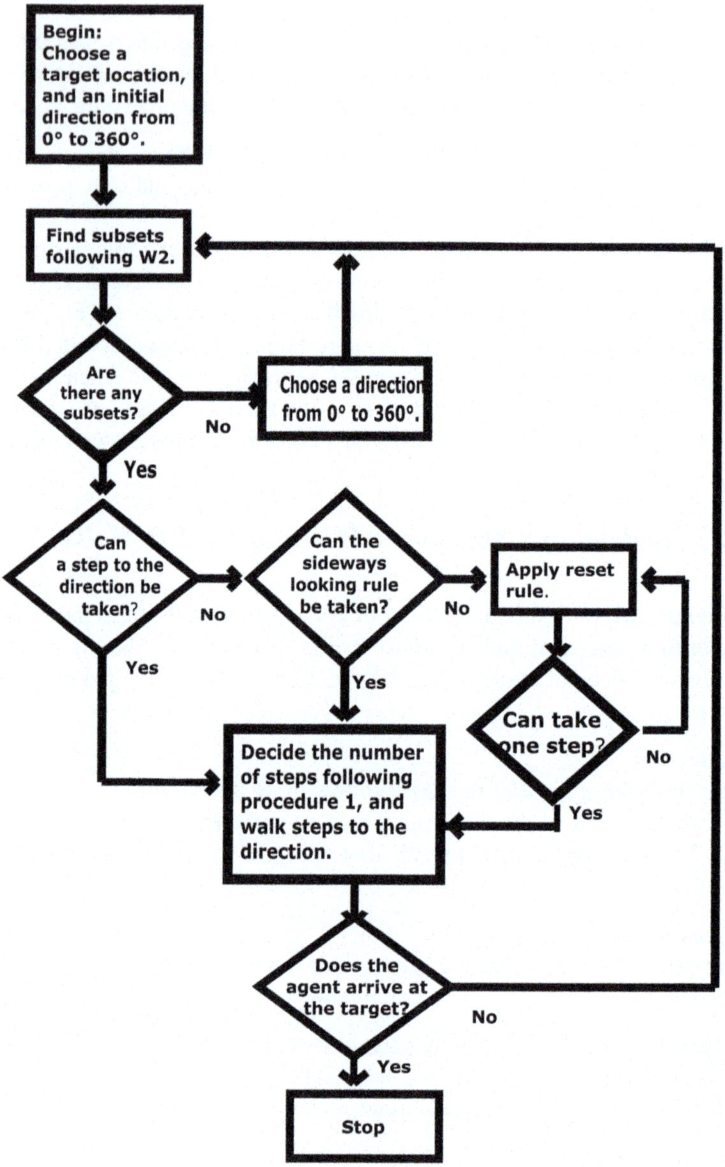

Fig. 10.5 Algorithm of the navigational movement model of the VD-Walker

Procedure 2

At each step, it is determined whether the target point and the number of straight movements have been reached. If so, the process proceeds to procedure 1. If not, procedure 2 is performed.

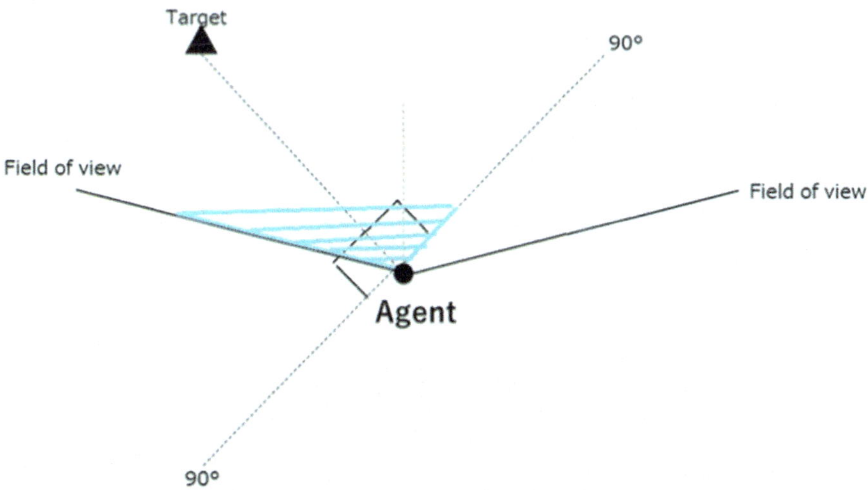

Fig. 10.6 Field of view of navigational movement

Procedure 3 (Irregular Case)
In the case of a wall collision, apply the sideways-looking rule (select 90° right or left direction). If the agent still cannot move, apply the reset rule (select 360° from a new direction) and execute procedure 1.

The wall vector field model (Lee 2015) also introduced a weighted origin destination model by limiting the field of view as the VD-Walker did.

In this navigational movement model, agents select the direction only from the overlapping area. In Fig. 10.6, only the angle marked with blue lines can be used to find the subset of bins. The area between the line that is 90° right and the line to the right of its field of view is excluded from the possible directions to choose.

10.4 Simulation Results

10.4.1 Basic Behavior of Agents

The VD-Walker agent's basic behavior is shown in Fig. 10.7. The simulation environment is a part of District K in Fig. 10.9. An agent starts from the edge of the white line, and the red circle is the agent with its lines of sight (yellow lines). The white line shows the trajectory of this agent.

An agent is set in an initial direction and following the algorithm, the agent chooses a direction and decides on a number of steps. When the agent moves the designated steps, it stops and again chooses a direction and number of steps. The agent repeats this again and again. In this simulation, the agent moves from back to forward.

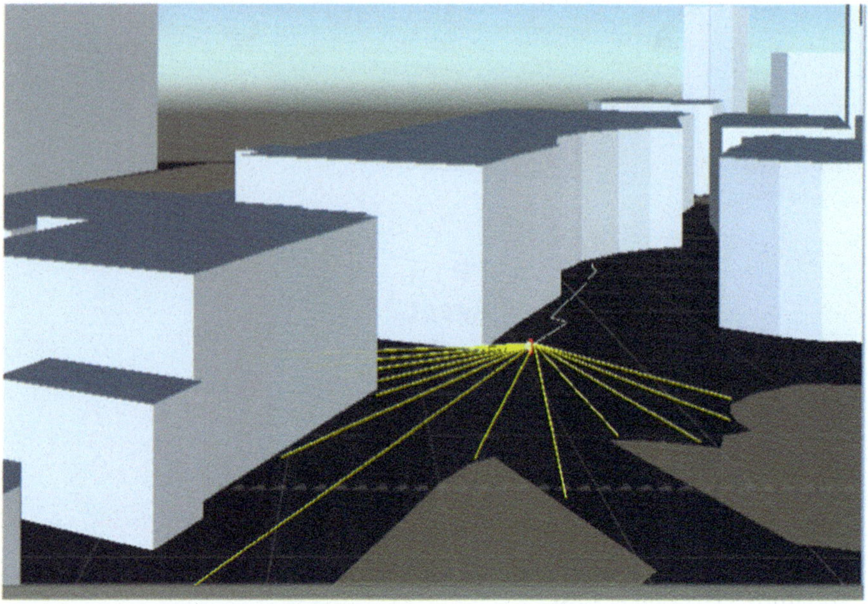

Fig. 10.7 Basic behavior of an agent

10.4.2 A Comparison of Agent Behavior When Other Agents Are Visible Obstacles or Not

The simulation environment here is the same as in Fig. 10.7 and this simulation result shows a view of the environment from above. There are four agents, shown as red dots, and some of the agents are shown with their lines of sight (yellow lines).

The lines of sight are only shown for choosing the next direction. The white lines show their trajectory. These four agents occurred from the same point.

When other agents are treated as visible obstacles, there is the possibility that the length of the line of sight may become shorter than the length of the walkable surface without the agents, because the length of the line of sight is the length to the agent. The length is used to choose the direction, and the agent makes a different selection. Agents keep space between them, and the tendency of straightforward movement becomes weak (Fig. 10.8).

10.4.3 Simulation in a Real District

Figure 10.9 shows simulation results in District K in Nagoya City. This district has a Central Japan Railway Company station and a subway station. There is a large-scale

(a)

(b)

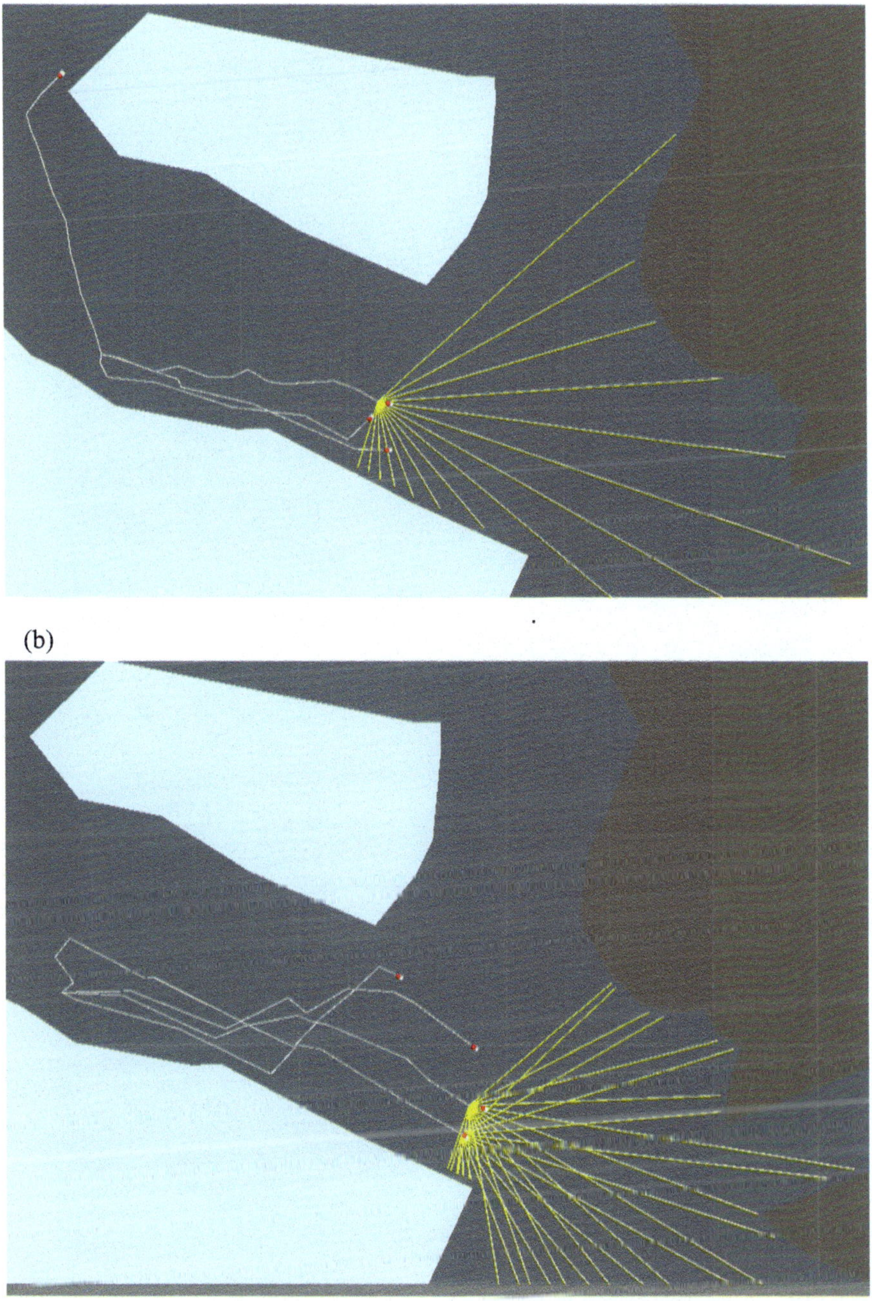

Fig. 10.8 Comparison of agent behavior when other agents are visible obstacles. (**a**) Agents are not visible obstacles to other agents. (**b**) Agents are visible obstacles to other agents

Fig. 10.9 Simulation in District K (Base case)

Table 10.2 The number of occurrences that occur from each occurrence point

Occurrence point	The number of occurrences	Occurrence interval (s)
1	46.2	65.1
2	32.2	94.1
3	69.4	43.2
4	71.0	49.2
5	57.8	51.9
6	926.8	3.2
7	290.3	10.3
8	516.4	5.8
Total	2000	–

commercial facility in front of the station where banks and hotels are located. Leaving the station, a civic center and park are there.

The numbers from one to eight in Fig. 10.9 correspond to the occurrence points in Table 10.2 The occurrence points are marked as black dots, and agents are marked as red dots. White dots show the trajectory of agents. In total, 2000 agents walked 1000 steps on average. The number of agents that occurred from each occurrence point is shown in Table 10.2. According to Table 10.2, 46.2 agents occurred every 65.1 s from occurrence point 1.

Figure 10.10 shows the simulation result in the same environment as Fig. 10.9 but this simulation shows the results when other agents are treated as visible obstacles.

When other agents are not treated as visible obstacles, agents can select their directions by probability that uses the walkable length precisely, and agents tend to move straight forward. Because agents cannot recognize other agents' existence, crashes may happen between agents. When other agents are treated as visible obstacles, agents behave in ways that allow them to avoid other agents. As a result, the tendency of moving straight forward becomes weak.

10.4.4 Navigational Movement Model of the VD-Walker

This simulation has the same eight occurrence points as the simulation shown by Fig. 10.9. Five agents occurred from each occurrence point at the same time. This simulation result shows that agents did not spread out but gathered at the target (Fig. 10.11).

Fig. 10.10 Simulation in District K (other agents are treated as visible obstacles)

Fig. 10.11 Simulation of navigational movement in District K

10.5 The VD-Walker on UNITY

10.5.1 A Suggestion from Gibson's Affordance and Vision-Driven Agents

Gibson's natural vision and affordance (Gibson 1978) and Hillier's natural movement (Hillier et al. 1993) are united in describing the behavior of pedestrian agents in the EVA model. Gibson's theory is about human visual perspective and deals with the usual space in which people live. The EVA model is a pedestrian model that is used only on a plate without height.

Turner and Penn (2002) referred to Gibson's concept of natural vision, which explained that pedestrians move from vista to vista. It is natural vision that is the topic authors would like to discuss in this book. The vision that could be checked by sitting in front of a machine in the laboratory was not Gibson's target. Gibson's research targets were ambient vision and ambulatory vison. Ambient vision refers to when the subject turns his or her head or looks around, and ambulatory vision refers to when the subject stands up or walks around.

Gibson explained that people learn about their environments, such as districts, when they are there, and when they have learned about their environment, their condition is referred to as orientation. People can learn a path to go to work, a path to go to the post office, a path to go to the grocery store, and a path to return home. When people can perform these behaviors in unknown districts, it is said that they have become oriented. Moreover, human could point the invisible place, where human had affordance.

Once the end of a path is reached, a new path appears. At the corner of the road, people can see the next road. This means that vistas are connected, and once vistas became clear by exploration, people can recognize whole environments such as houses, districts, and so on.

Not only people but also animals have the ability to move without hitting the edges, and this is a common characteristic of movements controlled by vision. The characteristics of a walking surface such as a path, water obstacle, cliff, step, or slope afford walking or afford collision. A cliff affords the inability to walk, and slopes can afford walkability or unwalkability depending on their angle or texture. People accept the limits of this 3D terrain.

Gibson's research is suggestive for the design of vision-driven agents. The VD-Walker model has the potential to expand 3D models as variations of the VD-Walker, using the 3D space model provided by UNITY.

10.5.2 Future Work of the VD-Walker

Though an EVA agent can pass another agent, in the current version of VD-Walker, an agent cannot pass another agent. An EVA agent is not treated as a visual obstacle, and neither is the VD-Walker's agent in the standard configuration. As a result, the VD-Walker's agent sometimes crashes. When comparing simulation results, it becomes clear that the behavior of the agents differs depending on their condition. The VD-Walker's agent has many parameters, including UNITY's parameters. It is necessary to consider the best way to use these parameters.

Because the VD-Walker is implemented on UNITY, it has the descriptive power of 3D, and 3D terrain models can be used. However, the current version of the VD-Walker does not use height information sufficiently. Applying the suggestive research of Gibson (1978) and considering a new version of the VD-Walker that fully uses height information and works on 3D terrain is important.

In our real world, there are multilevel pedestrian crossing decks from the exit of a station to downtown or elsewhere, as well as widely spread out underground shopping streets. In an urban district, spatial designs that utilize height differences are necessary. Pedestrian models that use natural movement and have explanatory power should be a research target, and a simulation system implementing those models is necessary for the use of a vision-driven agent simulation system as a tool for district design.

Acknowledgments This work was supported by JSPS KAKENHI GRANT NUMBER 18H03825.

The authors would like to thank Enago (www.enago.jp) for the English language review.

References

Architecture Institute Japan (AIJ) eds (1980) Data book of architectural design, vol. 3. Unit Space, Japan

Ferguson P, Friedrich E, Karimi K (2016) Origin-destination weighting in agent modelling for pedestrian movement forecasting. In: Proceedings: Eighth International space syntax symposium, pp 8152-1–8153-13

Gibson JJ (1978) The ecological approach to visual perception. Hough ton Mifflin, Boston

Hillier B, Penn A, Hanson J, Grajewski T, Xu J (1993) Natural movement: or, configuration and attraction in urban pedestrian movement. Environ Plann B Plann Des 20:29–66

Kaneda T, Maekawa Y, Kino T, Matsumoto Y, Susaki K (2019) Modeling and simulation of vision-driven pedestrian agent. In: Proceedings of the 12th space syntax symposium

Kaneda T, Ota A, Kobayashi J (2020) A study on factor analyses of pedestrians' spatial distribution inkanayama district, Nagoya-by applying Agent Analysis in Space Syntax studies. Transactions of AIJ. Journal of architecture, planning and environmental engineering. Edit Architectural Institute of Japan, Tokyo

Lee SJ (2015) Navigational pedestrian movement model with vision-driven agents. J Asian Architect Build Eng:371–378

Lee SJ, Lee KH, Kang SJ (2013) Study on a pedestrian simulation model of natural movement. J Asian Architect Build Eng 12(1):41–48

Penn A, Turner A (2002) Space syntax based agent simulation. In: Schreckenberg M, Sharma S (eds) Pedestrian and evacuation dynamics. Springer-Verlag, Berlin, pp 99–114

Turner A, Penn A (2002) Encoding natural movement as an agent based system: an investigation into human pedestrian behavior in the built environment. Environ Plann B Plann Des 29 (4):473–490

Chapter 11
Applying Vision-Driven Agent to Simulate Pedestrian Trajectory Under Furniture Layouts in Public Squares

Toshiyuki Kaneda, Yuri Morita, Satoshi Ohashi, and Takayuki Mizuno

Abstract In recent years, in Japan, public squares have been the focus of downtown revitalization, with the municipal sectors introducing tables, chairs, plantings, and stands that provide refreshments. In considering the efficient operation of public squares, it is important to analyze the influence of these layouts (hereinafter, layout) on pedestrian movement.

In this chapter, we investigate pedestrian movement in public squares where furniture layout is arranged, identifying vital characteristics in the process. We also attempt to simulate pedestrian-movement networks that differ according to layout by agent simulation, specifically Exosomatic Visual Architecture (EVA), which is used to represent the Natural Movement.

In this chapter, a pedestrian-trajectory survey is conducted in Toyota City Station Square. The results suggest that singularity examples deviating from the shortest-path principle tend to opt for the Longest "Line-Of-Sight" (LLOS) direction. Next, in addition to the LLOS principle, we explain the pedestrian-agent model, which has an OD (Origin-Destination)-tuple in the public square as well as an exit rule. The Toyama Grand Plaza, which is a rectangular square, is examined by matching the result of agent simulation by setting the condition of layout of furniture with the result of the trajectory survey. The simulation results for different layout settings are

T. Kaneda (✉)
Graduate School of Engineering, Nagoya Institute of Technology, Gokiso, Showa, Nagoya, Japan
e-mail: kaneda@nitech.ac.jp

Y. Morita
TOTO Ltd., Tokyo, Japan
e-mail: yuri.morita@jp.toto.com

S. Ohashi
Takenaka Corporation, Osaka, Japan
e-mail: ohashi.satoshi@takenaka.co.jp

T. Mizuno
Taisei Corporation, Tokyo, Japan
e-mail: mzntky00@pub.taisei.co.jp

T. Kaneda (ed.), *Downtown Dynamics*, Agent-Based Social Systems 16,
https://doi.org/10.1007/978-4-431-54901-7_11

also examined. Moreover, we also examine how to best handle complex spaces in a general sense.

11.1 Introduction

Municipal policies are currently dedicated to revitalizing downtown areas. Especially, the creation of liveliness in squares is attracting attention in downtown. In 2017, the National Institute of Land Infrastructure Management published a research report, in that such "lively squares" serve as bases for visitors' shopping activities and spread to the liveliness of the entire downtown. For the purpose of assessment of the location and the layout plan that suits the characteristics of the place, this report also examined the layout of chairs and plantings and analyzed visitor staying time and action range in some cases of the squares (Shinkai and Yoshida 2017).

Many researchers have examined public-space design; Hall (1966), Whyte (1980), Gehl (2010), and Gehl and Svarre (2013) are well known as contemporary survey styles; moreover, systematic open-data protocols (Gehl Institute 2017) have also recently appeared.

This creation of a "lively square" is interesting as a spatial-layout problem. When furniture is laid out functionally, the space would be separated into "staying zone" and "passage zone" clearly, for the former should be comfortable for visitors and the latter for passers. But on the other hand, furniture layouts are also intended to attract pedestrians to loiter in the zone, which is to say that spatial articulation should be vague.

Our pedestrian-agent simulation research has been prior used to address spatial-layout problems, with the models simulating spatial conflict between passers and detainees (Mizuno et al. 2016). Said research, in essence, examines the conflict between travel-queuing persons (backpackers) and commuter-transfer persons at Gold Clock square in Nagoya Station, which is a large transfer station. Herein, the travel-queuing agent is modeled to determine the "stay" position of pedestrians, which is explained by two main factors: the potential of a given spatial layout and the interpersonal distances from other "stay" agents. However, the commuter-transfer person is a simplified model that moves straight toward each waypoint while considering interpersonal distances.

In this chapter, the authors address former passers among the two kinds of agents appearing in the space-layout problem of a lively square using vision-driven agent simulation. Generally, pedestrian models that use vision mechanisms are referred to as vision-driven agents, such as exosomatic visual architecture (EVA) and Weighted Origin-Destination (WOD) in Chap. 10; however, here, it is a requirement that the passers have OD-tuples. Pedestrian models proposed in the urban and transportation studies of the 2000s belong to this category (i.e., Antonini et al. 2006; Asano et al. 2006). Our vision-driven agents have the greatest feature in action selection based on the longest line-of-sight (LLOS) principle.

This chapter reports the survey results concerning the actual state of pedestrian trajectory in squares where furniture is arranged and analyzes the characteristic features. In addition, pedestrian-movement lines that vary depending on the layout are reproduced through vision-driven agent simulations.

Note that this chapter deals with two-dimensional planes unless otherwise stated.

11.2 Fact Findings: Pedestrian Trajectory in Public Square

11.2.1 Toyota City Station Square

Toyota is a medium-size city of 300,000 or so residents in Aichi prefecture, with two stations located 300 m apart at its center: Toyota City Station of Nagoya Railroad and Shin Toyota Station of the Aichi Loop Railway. The square outside Toyota City Station is a second-level aboveground pedestrian deck connecting these two stations; the entrances of several large commercial facilities feed directly to the deck.

The deck was originally designed for efficient pedestrian passage, but, currently, as part of downtown revitalization and to create liveliness with temporary outdoor cafés with tables, chairs, and planting, the deck is asocial experiment, the aim of which is to encourage pedestrians to stay and use facilities in the square.

To obtain the characteristics of pedestrian trajectory, the Toyota City Station Square was selected to conduct a pedestrian-trajectory survey; Table 11.1 shows the survey outline and Fig. 11.1 the results.

11.2.2 Current Use of the Square

The surveyed pedestrian trajectories (OD-tuples) were summarized in an OD matrix, giving the following percentages: the flow connecting the Toyota City Station Ticket Gate 1 and Passageway 8 in the direction of Shin Toyota Station saw 36.4% of trajectories, Department Store 3 as the start/end point saw 26.5%, and Parking Lot 10 as the start/end point saw 12.4%. Indeed, the trajectory connecting Department Store 3 and 9 with Parking Lot 10 was often observed.

Table 11.1 Pedestrian trajectory survey in Toyota City Station Square

<table>
<tr><td>Target</td><td>Pedestrians passing through the survey target space</td><td rowspan="2"></td></tr>
<tr><td>Method</td><td>visually observe pedestrians traffic trace of 155 people randomly selected from the survey target.</td></tr>
<tr><td>Date</td><td>October 27 th, 2016 (Thu)</td><td rowspan="3">"Joyful Toyota Project" has been implemented for the purpose of creating daily hustle and bustle.</td></tr>
<tr><td>Time</td><td>A total of 1 hour 30 minutes from 16: 00 to 17: 30</td></tr>
<tr><td>Investigator</td><td>Record in the drawing with 3 people</td></tr>
</table>

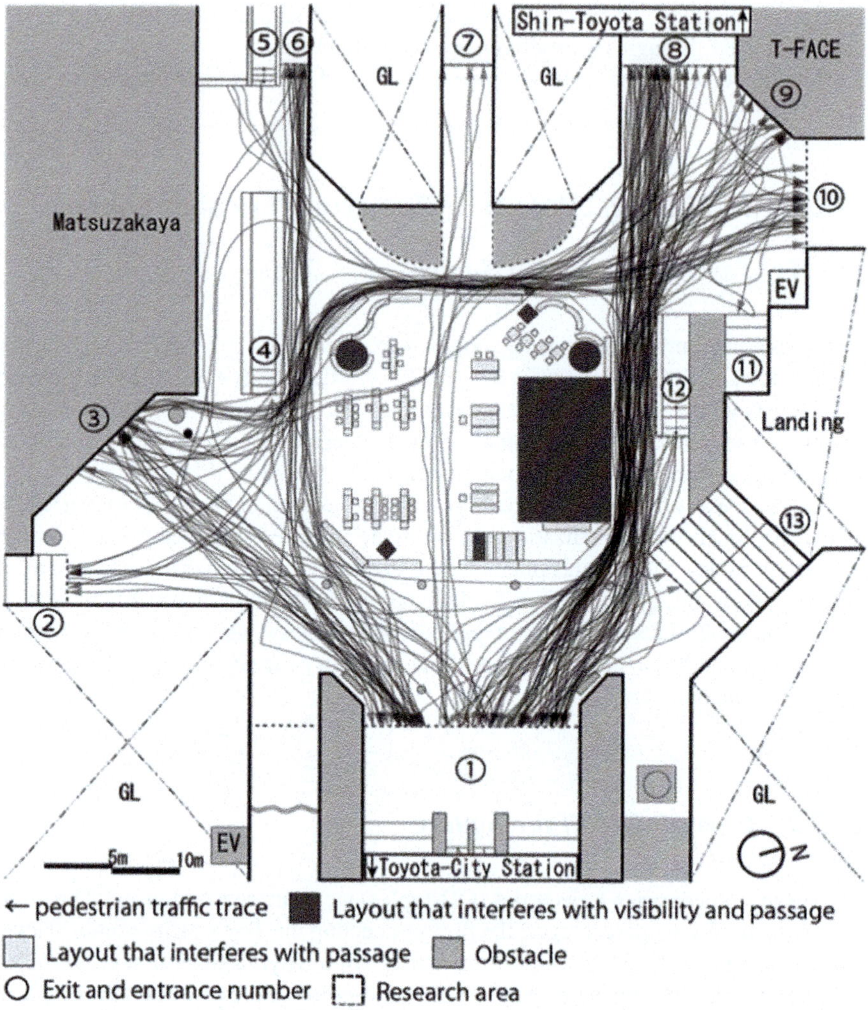

Fig. 11.1 Pedestrian-trajectory map of Toyota City Station Square

11.2.3 Analysis of Singular Trajectory Cases

Pedestrian-trajectory data with a redundancy rate of 15% or more was regarded as constituting singular cases; seven samples accounting for 4.52% of the entire trajectories were extracted, three of which are shown in Fig. 11.2. The redundancy rate is defined in Fig. 11.3.

In Case 1 with a pedestrian crossing from the southwest Stairway 5 to northeast Stairway 13, the trajectory through the square is the shortest path, but, due to the "oneness" of the layout, they avoided this path and moved in LLOS direction. Thereafter, when the destination was in sight, they changed direction and made for

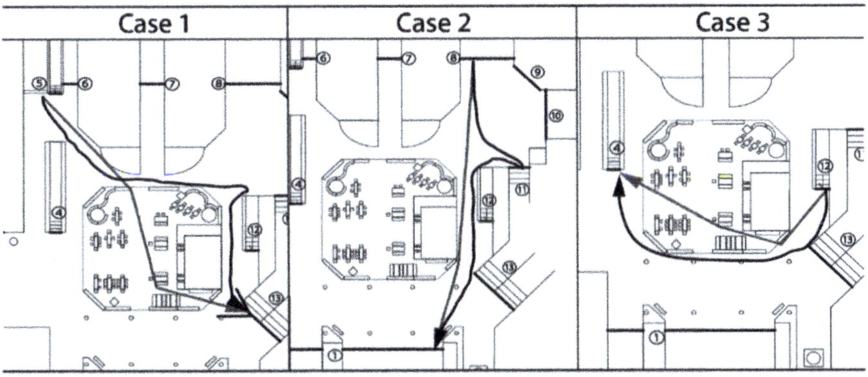

Fig. 11.2 Singular cases of pedestrian trajectory

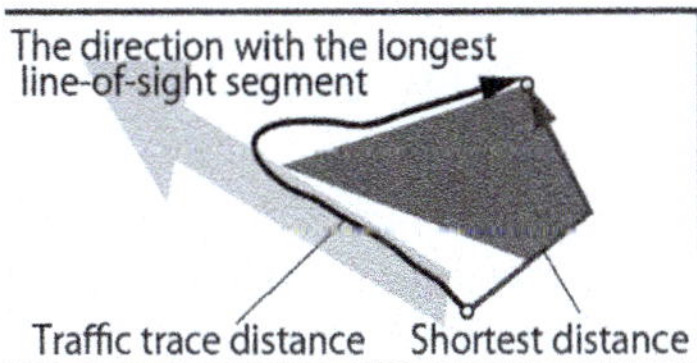

The redundancy rate (%)

$$= \frac{\text{Traffic trace distance} - \text{Shortest distance}}{\text{Traffic trace distance}} \times 100(\%)$$

Fig. 11.3 Redundancy rate of pedestrian trajectory

the exit, taking a short path. In Case 2, another pedestrian moves from Shin Toyota Station Passageway 8 to Toyota City Station Ticket Gate 1, first taking Stairway 11, which is the direction with the LLOS segment, but then changing direction for Toyota City Station Ticket Gate 1. In Case 3, a pedestrian walks from Stairway 12 (north) to Stairway 4 (south), taking the stairway up to the deck, they take a detour to avoid walking through the layout, which has a pattern similar to Case 1. From these results, it is evident that pedestrians with a high redundancy rate tend to move in the direction with the LLOS segment.

11.3 Vision-Driven Pedestrian Agent with OD-Tuple

11.3.1 Features of the Vision-Driven Pedestrian Agent

The Vision-Driven Pedestrian Agent (VDPA) acquires and acts on information within its vision field. Whereas conventional pedestrian agents are models in which personal spacing is added to the physics of fluids, gases, and particles, VPDA contains information about the outside world that is reflected in the vision mechanism; thus, it is a model in which spatial cognition affects behavior choice.

Agents whose external information in the vision mechanism is directly linked to micro-behavior selection are the pure vision-driven agent, and EVA by Turner and Penn (2002) became an epoch (detailed in Chap. 10). This chapter deals with a practical VDPA given destination information.

Here, the sight field is given a right and left visual angle limit with respect to the heading direction, and, in some cases, a visual distance upper limit is also given. Essential information in the sight field is the distance to the obstacle for each angle in a polar coordinate system with the heading direction as an axis.

There are many different types of vision mechanisms, but two in particular are important: (1) the direction-choice rule (which direction does the agent face?) and (2) the turning-condition rule (when does the agent turn?). As a pedestrian agent, VDPA continues straight until it satisfies the turning condition, performing direction-choice rules every time this condition is satisfied.

EVA (Penn and Turner 2002; Turner and Penn 2002) becomes one epoch for a short history of the VDPAs, which was an implementation of Hillier's natural movement in 2002. The Natural Movement means free exploratory behavior without a destination. In EVA, an agent moves 0.75 m per step on a two-dimensional plane with each step taking 0.5 s. Using a polar coordinate system, the basic rule of direction choice is to divide the sight field (visibility area) of 170° in front of the agent into equal-angled segments, determining the direction by using random numbers based on the probability in proportion to the bin area, which is proportional to the square of the sightline segment length as a representative of bin.

In addition, the turning condition in EVA consists of the fewest number of straight steps given stochastically (Poisson distribution). In the reported EVA, the direction is changed to the parameter, λ, after the number of steps (added to 1) is obtained by the probability based on the Poisson distribution given by 3. To summarize, EVA uses stochastic choices for the direction choice and turning condition.

Ferguson et al. (2012) modified with EVA by adding a destination factor, ultimately developing a version called the Weighted Origin-Destination (WOD), which limits the sight field to an angle consisting of a point where the distance to the destination is shortened (Ferguson et al. 2012; Uyars et al. 2017). This is referred to as a VDPA with destination information. The details of EVA and WOD are mentioned in Chap. 10 by Susaki and Kaneda.

11.3.2 Existing Vision-Driven Pedestrian Agents

For VDPA, where destination information is given exogenously, a wide variety of technical methods are conceivable depending on internal configurations as well as the balance between free exploration orientation and destination orientation.

In modeling vision-driven agents with destination information, a number of different technical methods can be conceived with respect to the significant manner of the internal compositions of pedestrian agents, as well as the balance between an agent's free exploration orientation and destination directionality.

In direction choice, there are stochastic and deterministic selection types that determine the next direction of movement from information on position and direction. The same can be said for the turning condition, with stochastic and deterministic selection types based on distance, time, and position.

The LOS agent by Turner et al. (2005) is a variant of EVA. Some changes have been made where the viewing center axis is always set in the direction toward the destination; moreover, a probability proportional to the line segment length is used for direction choice.

In the wall vector field model established by Lee et al. (2013) (see also Lee 2015), direction choice is defined as the synthesis (integral) of sightline vectors in the sight field, wherein the sightline vector in the polar coordinate system is multiplied by a correction coefficient called the attractive factor. In addition, the turning condition is applied at every step.

LSEM (Kaneda et al. 2017) selects the direction of the longest-sight line segment in the sight field, which is deterministic; however, the turning condition is set for every step (0.5 s, 0.4 3 m in distance) by stochastically using a probability 1/3 random number.

For pedestrian agents with a given destination, Antonini et al. (2006) proposed a mixed nested logit model that selects a bin with a velocity vector (direction and velocity) divided into 11 segments in a 170° sight field every 0.9 s. This research also estimated related parameters using video data.

As a vision-driven agent that deals with congested situations, Asano et al. (2006) proposed a model that determines the velocity vector based on the principle of minimizing the direction-deviation cost and obstacle-approach cost, in which the movement of other people was considered.

The agent model, VDPA1, in this research is based on the LLOS model with terminal navigation; when an agent finds their destination, the agent moves straight to their destination.

11.4 Pedestrian-Agent Simulations Based on the Trajectory Survey

11.4.1 VDPA1: Our Prototype Version of the VDPA

To conduct simulations, this research selected Toyama Grand Plaza as the spatial case; the plaza has a simpler spatial configuration compared with the Toyota City Station Square described later.

Toyama Grand Plaza, located in the medium-scale city of Toyama, is a rectangular all-weather atrium measuring 65 m from north to south and 21 m from east to west. As part of downtown revitalization, two urban redevelopment projects have

Table 11.2 Pedestrian-trajectory survey in Toyama Grand Plaza

Target	Pedestrians passing through the survey target space
Method	visually observe pedestrians traffic trace of 159 people randomly selected from the survey target.
Date	November 24 th, 2016 (Thu)
Time	A total of 1 hour 30 minutes from 13: 30 to 15: 00
Investigator	Record in the drawing with 3 people

It is a square where two commercial facilities and shopping districts are connected by a glass roof. There is an elevating stage and large vision.

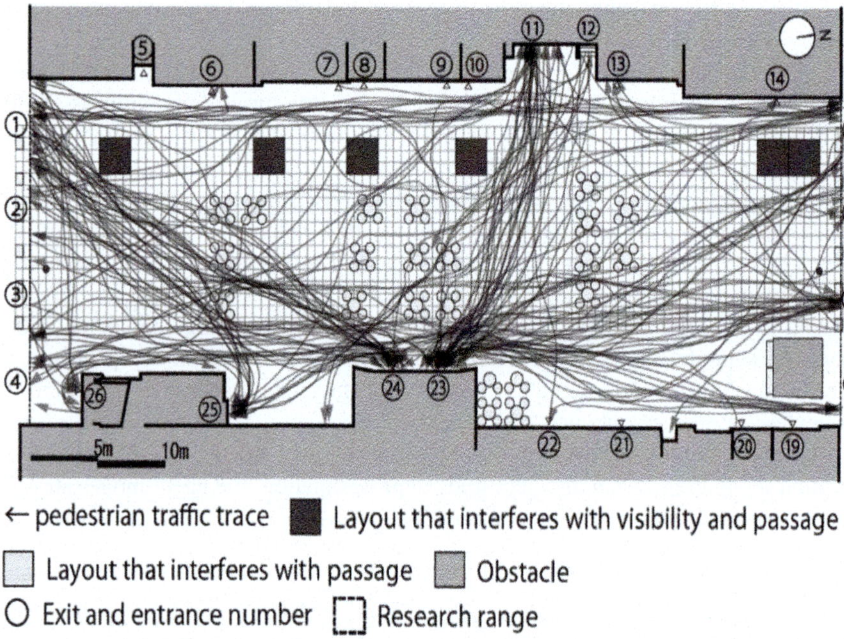

Fig. 11.4 Pedestrian-trajectory map of Toyama Grand Plaza

been carried out; by replacing public open grounds and streets, the plaza was created in 2007 on the ground section. A large department store is located on the west side and a public parking-lot building with ground-level shops on the east side.

To attract customers, an elevating stage was designed along with a layout of temporary tables, chairs, planting, etc., and, as a multipurpose space, 100 or so events are held annually.

As shown in Table 11.2, this research conducted a pedestrian-trajectory survey at Toyama Grand Plaza. Figure 11.4 shows the survey results.

11.4.1.1 Algorithm of VDPA1

VDPA1, in which the LLOS model and terminal navigation were added to the EVA, was used for simulations. The terminal navigation mentioned here is restricted by the rule that, when an agent finds their destination, they head straight for it. The initial value of the agent's movement direction is randomly set within a range of 30° to the right and left with respect to the given destination direction. Figure 11.5 shows the behavior algorithm for VDPA1. The vision-field angle for VDPA1 was set as 168.75°; the number of the bins was set as 60.

11.4.2 Validation of VDPA1: Comparing Survey Data and Simulation Results

Firstly, to validate VDPA1, the same layout as that in Fig. 11.4 was created on a simulator. Each entrance/exit position was set based on the survey data, as well as the number of agents appearing/leaving at each entrance/exit. Figure 11.6 shows the trajectories of the simulation results, from which it is evident that the survey results have the highest number of trajectories passing through the inside of the layout, especially between the west-side End 1 of the south side of the plaza and Entrance 24 of the large department store on the east side, followed by the shortest-path trajectories passing between Entrance/Exit 11 and 12 to the parking lot on the west side and Entrance 23 of the large commercial facility on the east side. Indeed, the

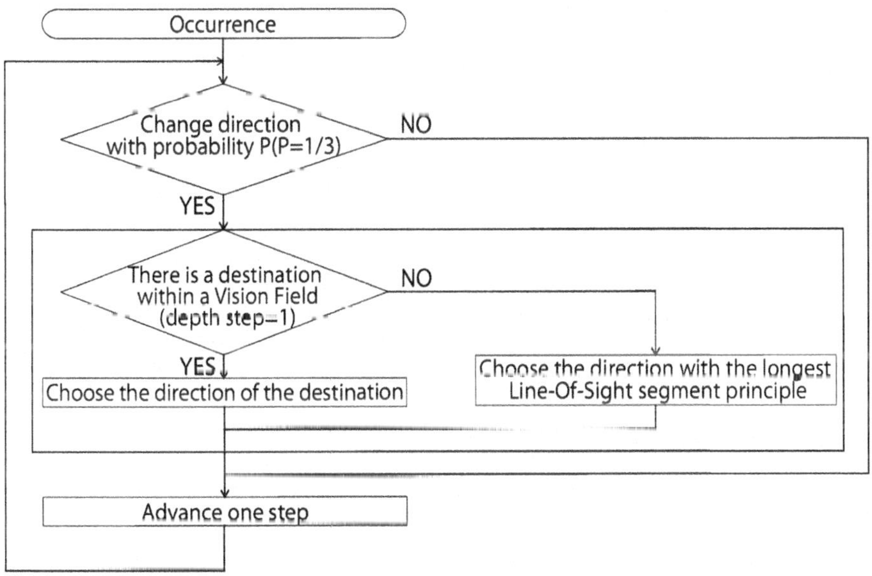

Fig. 11.5 Spatial-behavior algorithm in VDPA1

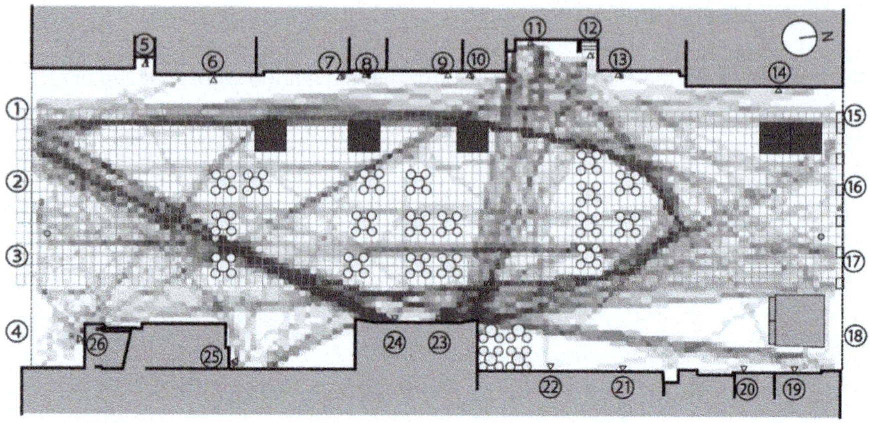

Fig. 11.6 Simulated trajectories (Basic Case)

simulation results have similar tendencies. In particular, similarities can be seen with the trajectories between Entrance/Exit 3 and 17 (north and south) and trajectories obliquely passing through the inside of the layout between south side Entrance/Exit 4 and Parking Lot 11 and the north-side Entrance/Exit 17 and Entrance 23 of the department store.

The survey results show a tendency where all the trajectories across north and south, specifically between Entrance/Exit 1 and 2 to 17 and 18, pass through the east side of the plaza, whereas, in the simulation results, the agents passing from Entrance/Exit 1 and 2 to 17 and 18 (south to north) typically move through the west side of the plaza. This is because, when an agent's field of sight was blocked by the layout, they opted for the direction with the LLOS segment. Accordingly, trajectories taking a detour northwest of the layout are only seen in Fig. 11.6, which consists of the simulation results. This is the most noticeable difference between the agent trajectories in VDPA1 and the actual pedestrian trajectory. Despite this, the model has a favorable goodness of fit.

11.4.3 Simulations with Different Furniture-Layout Settings

In comparison with the basic case shown in Fig. 11.6, simulations were conducted with the two different layouts shown in Figs. 11.7 and 11.8. In Case 1, a tight layout of furniture pieces (9 or 10 tables, and 1 or 2 planting) was set at two locations in the plaza: one northwest and the other southeast. In Case 2, eight tables and four plantings were set on the north side of the plaza and an event space was set on the south side. Compared with Fig. 11.6 (basic case), Case 1 had fewer agents passing through the layout, which indicates the strong "oneness" of the design. In addition, the trajectories of the agents crossing between north and south (Entrance/exit 1, 2 and 17, 18) moved to the east side. In Case 2, we can see an increase in the number of

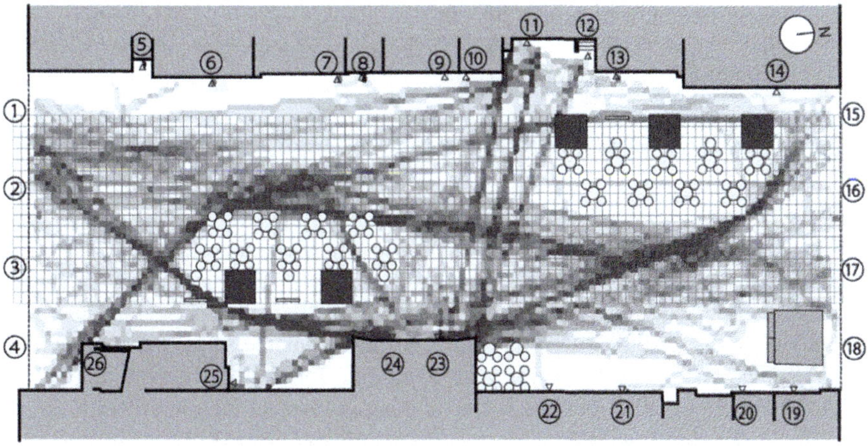

Fig. 11.7 Simulated trajectories (Case1)

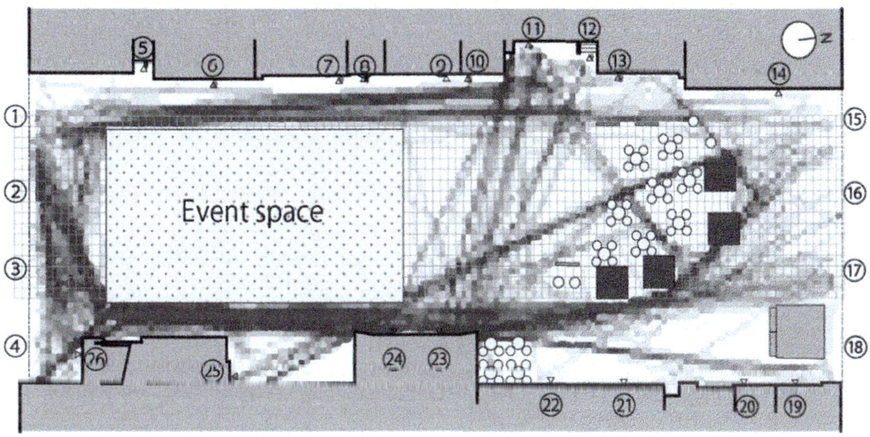

Fig. 11.8 Simulated trajectories (Case2)

trajectories passing through the east side of the event space and a reduction in the number of trajectories passing through the west side. Moreover, many agents can be seen to cut through the layout.

11.5 Conclusion

In this chapter, a pedestrian-trajectory survey is conducted at the Toyota City Station Square and the characteristics of pedestrian trajectory are analyzed. Thereafter, the VDPA is reviewed with respect to a two-dimensional plane. In particular, the

VDPA1 is discussed in which terminal navigation was added to the LLOS principle. The simulation results and pedestrian-trajectory survey data were compared with the Toyama Grand Plaza to examine any differences.

1. *Driven by straight forward line*
   In this research, following the traditions of Hillier's natural movement in his space-syntax theory, the LLOS segment is represented as a visual line. In reality, the LLOS segment is the projection of a visual line in a three-dimensional space onto a two-dimensional space, which is a highly abstract concept.
   In VDPA1, a vision field is composed of lines that allow straight forward movement. Please note that, even though an object can be seen beyond chairs and tables, in this chapter it was treated as not being visible to the agent. In reality, tables and chairs constitute obstacles to straight forward movement, but they do not block visual lines beyond them. In contrast, typical planting/facilities physically block the straight forward movement and the visual line.
2. *Space handled by VDPA1*
   To some extent, VDPA1 was successful in reproducing pedestrian trajectory in the Toyama Grand Plaza, but the same cannot be said for Toyota City Station Square.
   Under the LLOS principle, in the case of a rectangular square (generally, a convex polygon), if there is no obstacle, this function forms a one-peak shape, at which point, even with the presence of minor obstacles, the agent is able to identify the destination without considerable deviations. Indeed, the obtained trajectories are consistent with the survey results. However, in the case of a more complex-shaped square, this function has a multi-peak shape, and the agents separately move forward in multiple directions, each of which is locally the longest; for this reason, many agents wander around without being able to find their destination. Accordingly, this model needs to be further improved to handle more complex spaces.
3. *Sense of "oneness": The possibility of space assessment by simulating the integration of recognition behavior*
   The layout of furniture in a square, such as tables, chairs, and plantings, is discussed in the field of architectural planning in terms of cognitive psychology, especially their effect on the degree of "oneness" or the sense of "oneness." In contrast, behavioral science approaches to observe whether or not to actually pass through the layout provided are being developed rapidly. The case of Toyama Plaza mentioned at the beginning is that there is abundant experimental knowledge to fit the proposed model (Shinkai and Yoshida 2017: 60). The vision-driven agent simulation discussed in this chapter offers an integration of both approaches. Indeed, it is expected to provide a valuable contribution.

   With the research results of Chap. 9 and this chapter, the authors have decided to explore our research program for their practical applications of vision-driven agent simulations. This involves the data collection of micro-behavior using virtual space in Chap. 12, which provides a strong foundation for vision-driven agent simulations. In addition, this also includes the subsequent VD18s model, which is a vision-driven agent that handles related sign layouts to facilitate trajectory identification by visual recognition in Chap. 13.

References

Antonini G, Bierlaire M, Weber M (2006) Discrete choice models of pedestrian walking behavior. Transp Res B Methodol 40(8):667–687

Asano M, Kuwahara M, Tanaka S (2006) Micro pedestrian behavior model in congested conditions. In: 5th ITS symposium, pp 418–424

Ferguson P, Friedrich E, Karimi (2012) Original situation weighting in agent modelling for pedestrian movement forecasting. In: 8th International Space Syntax Symposium, pp 8153.001–8153.013

Gehl J (2010) Cities for people. Island Press, Washington, DC

Gehl Institute (2017) The open public life Data Protocol version: beta. https://gehlinstitute.org/tool/public-life-data-protocol/. Accessed 27 Sep 2017

Gehl J, Svarre (2013) How to study public life. Island Press, Washington, DC

Hall E (1966) The hidden dimension. Doubleday, Garden City, NY

Kaneda T, Okamoto K, Suzuki T, Tamada M (2017) An analysis of the characteristics of a proactive exploration behavior model based on three-dimensional virtual space experiments. In: Putro US, Ichikawa I, Siallagan M (eds) Agent-abased approach in economic and social complex systems IX. Springer, Singapore, pp 191–197

Lee SJ (2015) Navigational pedestrian movement model with vision-driven agents. J Asian Architect Build Eng 14(2):371–378

Lee SJ, Lee KH, Kang SJ (2013) Study on a pedestrian simulation model of natural movement. J Asian Architect Build Eng 12(1):41–48

Mizuno T, Shomitsu M, Kaneda T (2016) A study on the agent-simulation of waiting behaviors inside a transfer station 2016. In: IEEE 40th annual international computers, software and applications conferences, pp 111–116

Penn A, Turner A (2002) Space syntax-based agent simulation. In: Schreckenberg M, Sharma SD (eds) Pedestrian and evacuation dynamics. Springer-Verlag, Berlin, pp 99–114

Shinkai H, Yoshida J (2017) A study on evaluation of urban public squares: field survey-based examination of procedures to evaluate the quality of public space. In: Technical Note of National Institute of Land Infrastructure Management, No. 971

Turner A, Penn A (2002) Encoding natural movement as an agent-based system: an investigation into human pedestrian behavior in the built environment. Environ Plann B Plann Des 29 (4):473–490

Turner A, Mottram C, Penn A (2005) Line-of-sight pedestrian microsimulation in urban systems. In: 9th international conference on computers in urban planning and urban management, pp 210.001–210-010

Uyar G et al (2017) Using an analytical process to contextualise architectural prototypes for metro stations. In: 11th international space syntax symposium, SSS 2017, pp 118.1–118.16

Whyte W (1980) The social life of small urban spaces. Conversation Foundation, Washington, DC

Chapter 12
A Virtual Space Gaming Experiment System for Analyzing Waiting Behavior in a Public Square

Tomohiko Kino, Satoshi Ohashi, and Toshiyuki Kaneda

Abstract The aim of this study is to discover the characteristics of waiting behavior in a public square and the influence of others on waiting behavior. We develop a virtual space gaming experiment system that simulates waiting behavior in public spaces. This system was based on square research references and a behavioral experiment in the real space. The system was given the additional feature of eyeball expression, which simulates an awareness of other people's gazes. In addition, the research clarifies the types and characteristics of waiting behaviors in gaming experiments. First, we categorized exploration behaviors to determine waiting spots and the factors that determine the body direction at a waiting spot. Second, the selection of a waiting spot can be categorized using three conditions relating to visual recognition by others, which are eye contact, gaze, and personal space. It was found that there is a relaxation order on these conditions when increasing the crowd density.

12.1 Introduction

12.1.1 Research Background and Objectives

In recent years, public squares across Japan have been included in many developments to revitalize urban life. These squares form the arena for several types of behavior, such as pedestrians passing through or using the facilities within the square

T. Kino
Taiseki Corporation, Tokyo, Japan
e-mail: kn-tmh00@pub.taisei.co.jp

S. Ohashi
Takenaka Corporation, Osaka, Japan
e-mail: ohashi.satoshi@takenaka.co.jp

T. Kaneda (✉)
Graduate School of Engineering, Nagoya Institute of Technology, Gokiso, Showa, Nagoya, Japan
e-mail: kaneda@nitech.ac.jp

T. Kaneda (ed.), *Downtown Dynamics*, Agent-Based Social Systems 16,
https://doi.org/10.1007/978-4-431-54901-7_12

and those waiting for a meeting or appointment. Waiting is a common daily activity for those familiar with a square as well as strangers to the area; by analyzing the characteristics of waiting behavior, it is anticipated that information useful for the planning and designing of public squares will be revealed. The present research also pays attention to the degree to which the presence of other people within a square has an influence on an individual's behavior. It is reasonable to assume that some factors, particularly personal space and line of sight, will be affected by the presence of other people and will substantially influence the behavior of a person who is waiting.

The purpose of this research is to develop a virtual space gaming experiment system with a consideration of the influence of other people and to analyze the characteristics of waiting behavior within a public square through virtual space experiments using the aforementioned system.

Among the varieties of meeting behavior, this research analyzes the "waiting behavior" of the first person to arrive at a meeting spot. Their waiting behavior is analyzed from three aspects: exploration behavior in relation to looking for a good spot within the square to wait for the other person; decision behavior with respect to determining the location of the waiting spot; and decision behavior in relation to determining the body direction during waiting.

12.1.2 Classification of Past Research and Positioning of This Research

In the field of architectural planning, Yoshitomi and Nakamura investigated meeting behavior in a public square (1996) and distinguished between the first-comer's waiting behavior and the second-comer's "exploration behavior" to find the first-comer; however, the research concentrated on the second-comer's exploration behavior, and the first-comer's waiting behavior was not fully analyzed. Tsuchida et al. (2005) conducted an analysis by using the place-marking method that essentially plots waiting spots on a plan, but the existence of other people, i.e., crowd density, was excluded.

Seta et al. (1997) developed a visual simulator (space experiment system) in which sensory behaviors were directly linked with screen images, thereby enabling real-time simulation along with a sense of reality. Yasufuku et al. (2006) developed an evacuation simulator enabling virtual space gaming experiments within a virtual building and analyzed evacuation behavior at the time of flooding in an underground space. Furthermore, Suzuki et al. (2015) and Kaneda et al. (2015) analyzed active exploration behavior by using virtual space experiment systems and applied the results to agent simulations. All these previous studies indicate the possibility of expanded research using an experiment system. In addition, VDPA1 (see Chap. 11) assessed the influence of the layout of furniture in open spaces on pedestrian movement through agent simulation.

Our research has developed a virtual space experiment system to simulate waiting behavior within a public square. Moreover, to study the influence of other people on the behavior of first-comers, we have developed an effective way to express other people. We consider that conducting virtual space experiments using the virtual space experiment system and post-analysis with a focus on relationships with other people make it possible to clarify the relationship between the existence of others and the type of waiting behavior in a square. In addition, by extracting a human agent behavior model using the experiment system, the implementation of a highly practical agent simulation of different crowd densities in a square can be expected.

12.2 Extraction of Distances from Other People Based on Square Research References

This section explains distance relationships with other people, extracted from past research (Fig. 12.1). This chapter focuses on two types of distance that are considered important in the recognition of other people, the first of which is 25 m, as shown by (4). According to Alexander (1977), within this distance, a person can recognize other people's facial features; therefore, it is considered that a person can be identified. In our paper, this distance is referred to as the "identification distance." The second distance is 5 m, as shown by (6). Hall (1966) explained that within this distance, eye movements are clearly visible, so a person will feel under observation and will be aware of other people both physically and psychologically. In our chapter, this distance is referred to as the "eye contact distance."

Distance to other [m]	Distance number	Reference number	Distance type	Explanation of distances
500	①	A	limit of identification	Humans can be recognized as different from animals, shrubs, rocks, etc.
300 100	②	A	limt of body language	Big movements and body language are recognized. Individuals can be recognized, but not their physical features.
70 50	③	A	limit of recognition of physics characteristics	Individuals can be identified from physical features such as body shape and gait. Facial features are not clearly visible.
25	④	A.B	Limit of identification of facial features	Facial features are clearly recognized and individuals identified, but eye contact is not possible.
22 7.6	⑤	C	Limit of personal space	Personal space: within this distance, a person feels uncomfortable with a stranger.
5.0	⑥	C	Limit of recognition of eye contact	Eyes and facial features of others are visible within the visible fovea(Note 1). Individuals begin to interact with each other with eye contact.
3.7 [m]	⑦	C	limit of approaching strangers	Individuals have a strong awareness of others, and to prevent social awkwardness, avoid strangers.

A:Cities for people [8] B:A Pattern Language [9] C:The Hidden Dimension [10]

Fig. 12.1 Extraction of distances to others based on references

12.3 An Experiment on a Real Space

Figure 12.2 provides an outline of the real space experiment.

With respect to the survey purpose, as a preliminary experiment to develop a virtual space experiment system, we conducted an experiment concerning "(1) Waiting behavior." To verify the importance of the two types of distance explained in the previous section, we also conducted an experiment concerning "(2) Degree of awareness of others."

The survey site was the Otagawa Station West Plaza in Tokai City, Aichi Prefecture. This plaza had been part of a station redevelopment project. Weekend events and other activities are held to create a lively environment. An area of approximately 2800 m^2 within the square was set as the survey observation area.

1. Waiting behavior
   Subjects were instructed to enter the experiment area and walk freely until they had decided on what they considered to be the best spot to wait. The experimenters recorded their behavior and interviewed the subjects after their decision had been made. The experiment conditions are shown in Fig. 12.2.

   Figure 12.3 shows the positions of waiting spots and the direction of the body at each spot, and Table 12.1 outlines their reasons for deciding on the spot and the direction. At the waiting spots, the following behaviors were observed: sitting on chairs at the center of the square, which was close to the station and allowed a clear view of the whole area, or standing while leaning on an information board or pillar. The body direction at a waiting spot was determined by concerns regarding the presence of other people or the awareness of eye contact (more specifically, directions in which they could easily see people's movements or avoid eye contact). It seems that in waiting behavior, subjects were aware of the possibility

Subject	Five students (M 4, F 1)	Survey area	West Plaza at Otagawa Station in Japan
(1).Waiting behavior	Instruct subjects to walk to a spot they feel is most suitable for waiting in an open space (right photo), and record the spot and the trace. Then interview them about their actions.	Photo	
(Situation)	① You are meeting outside commuting hours ② You will arrive 15m. early and wait about 15 minutes. ③ You do not know from which direction the other person will approach from. ④ Do not use a phone, etc. to help you meet. ⑤ No one else is in the square.		
(2).Degree of awareness of others	Place another person on the numbers shown on the right (4 m, 20 m, 35 m) and have them face and face away from the subject. Evaluate using the SD method items below and record their answers.	Schematic diagram of measuring position	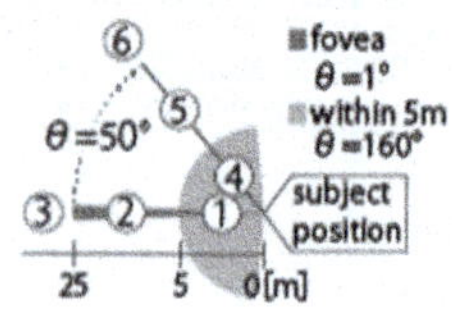
(Question)	Q1 Can you see the direction the person is looking in? Q2 Does it concern you? Q3 Can you identify their facial features? Q4 Can you identify them? Q5 Are you aware of other people?		

Fig. 12.2 Outline of real space experiment

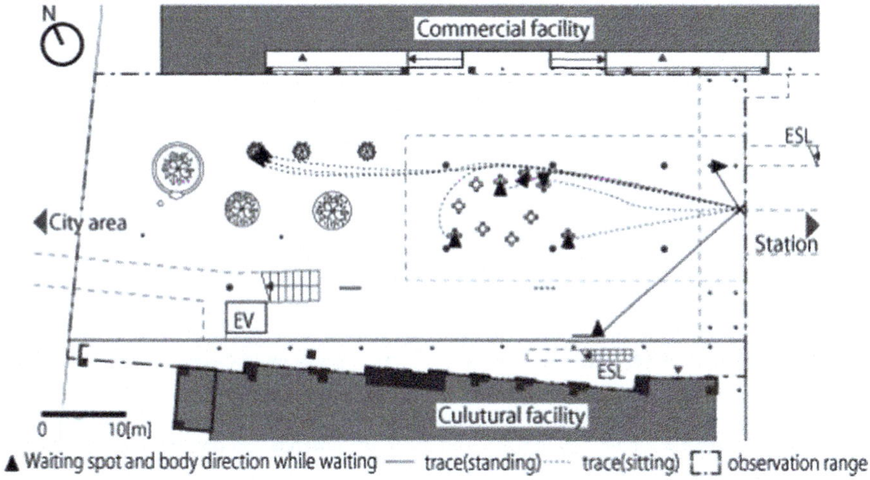

Fig. 12.3 Waiting spots and body directions while waiting ($N = 10$, two trials for five people)

Table 12.1 Reasons for determining the waiting spot and body direction while waiting ($N = 10$, two trials for five people)

Waiting spot	• It is close to the station	• I wanted to sit down
	• It is empty of people	• There is a clear view
Body direction while waiting	• I can see the whole area by turning my head	
	• I can see the commercial facilities	
	• It is easy to see people's movements	
	• I want to avoid eye contact with others	

of visual recognition by other people and engaged in behavior to keep a comfortable distance from them.

2. Degree of awareness of others
Other people were positioned on the numbered spots shown in the schematic diagram at the lower right of Fig. 12.2 (facing and not facing the spots of 4 m, 20 m, and 35 m). The subjects were instructed to answer questions according to the SD method evaluation items.

Figure 12.4 shows the relationships between the degree of awareness of others and the distance from them. The value of each item is the mean value of answers. Question 2—"Does it concern you?"—received positive answers for both facing and not facing at a distance of 4 m. Question 4—"Can you identify them?"—also received positive answers for both facing and not facing at a distance of 20 m. These results support the importance of two types of distance: "identification distance" and "eye contact distance." The results also confirmed that the not facing positions had a lower degree of awareness of others than the facing positions, indicating a tendency for subjects to feel comfortable.

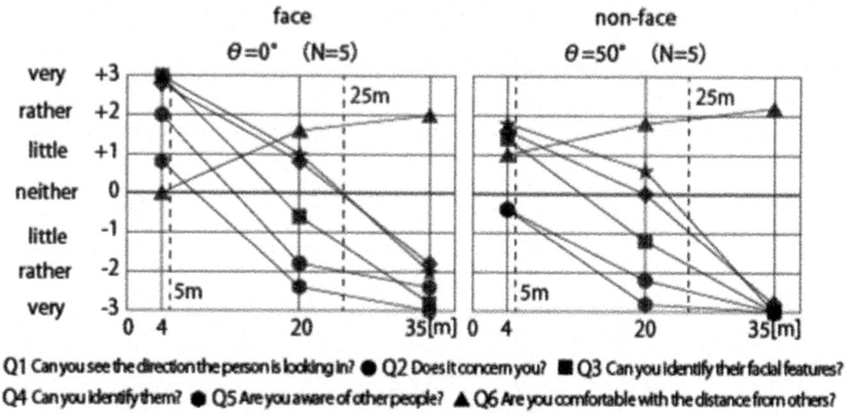

Fig. 12.4 Degree of awareness of others according to the difference of distances (the mean)

12.4 Gaming Experiments Using a Virtual Space System

12.4.1 Structure of the Virtual Space System

Using artisoc 4.0, a virtual space experiment system was developed (Kozo Keikaku Engineering 2020). Figure 12.5 shows the operation procedure.

To understand the characteristics of waiting behavior and waiting spots, two virtual squares with different layouts of pillars and walls were prepared and used for the experiments described later (Fig. 12.6).

In addition, to examine the influence of other people on waiting behavior, with reference to past research, three types of population density were prepared: low

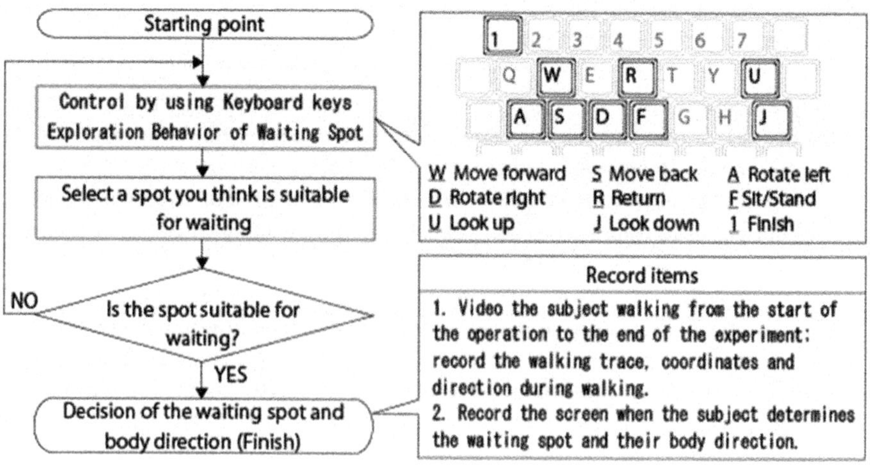

Fig. 12.5 Operation of virtual space experiment system

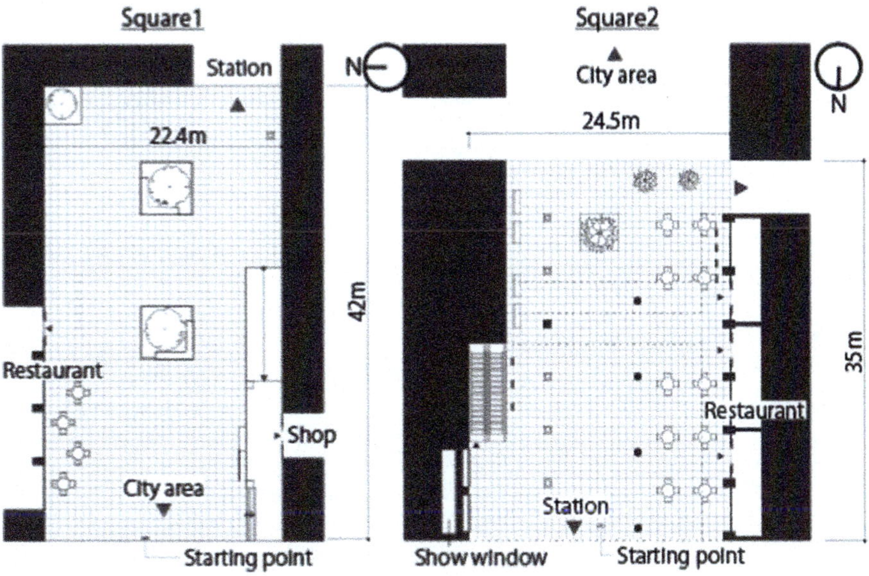

Fig. 12.6 Floor plans of virtual squares

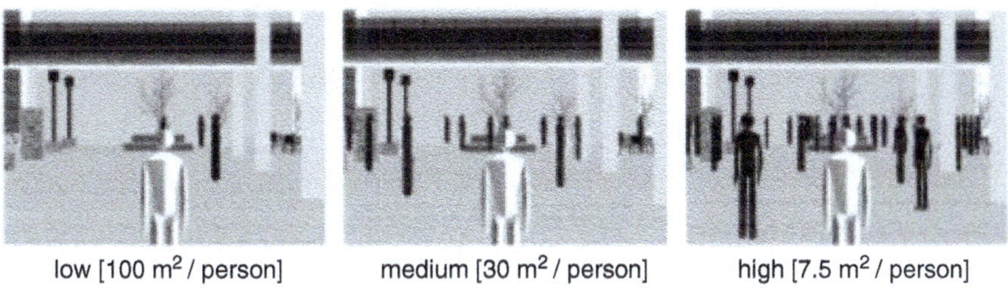

Fig. 12.7 Different views according to population density (Square 2)

density (100 m^2/person), medium density (30 m^2/person), and high density (7.5 m^2/person) (Fig. 12.7).

Furthermore, since former systems had had difficulties in incorporating the awareness of distance to others, other agents were introduced; their eyeball size changed according to the two thresholds of eye contact distance and identification distance (Fig. 12.8). In addition, all of these other agents did not move.

12.4.2 Gaming Experiments Using the Virtual Space System

Table 12.2 provides an outline of the gaming experiments. An experiment using the virtual space experiment system and involving 15 engineering department students

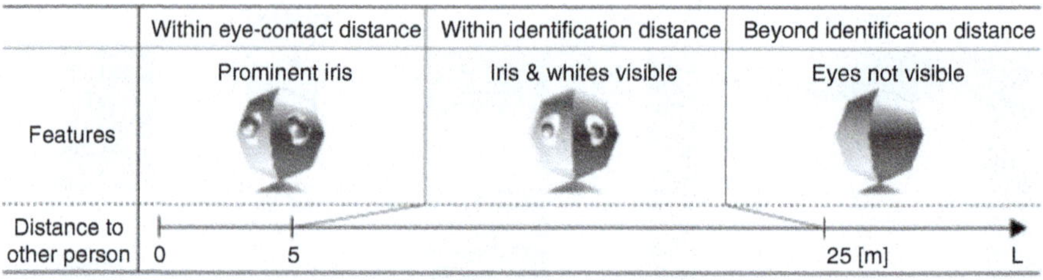

Fig. 12.8 Features of others according to distance

Table 12.2 Outline of the gaming experiments

Subject	15 students (M11, F4)	Time per case	15 min
Procedure	1. For the subject to recognize the virtual square, have them look around and understand the square layout, and then determine their start direction		
	2. Walk through the square taking any route to the spot they think is most suitable for waiting		
	3. After the experiment is completed, ask the reasons for their choice		
	4. Ask for the reasons why at a particular moment the subject suddenly changed direction, etc. both at the time just before starting to walk and during walking		
Scenario and instructions to subjects	• It is your first visit to this square		
	• It is not a holiday or during commuting hours		
	• You have arranged to meet a close friend		
	• You have arrived early and will wait for about 15 min		
	• You do not know which direction your friend will come from		
	• There is no need to consider the weather, season, or time of day		
Case	• Square (square 1: an outdoor station square, square 2: an indoor station square)		
	• Density (low medium high)		
	Total of 6 cases = 2 squares × 3 densities		

was conducted and videoed. In this experiment, each subject looked out across the whole square before walking and standing upon a spot that they considered the most suitable waiting spot. Subsequently, the students decided on the direction of their body and then pressed a button to finish. At the end of the experiment, each subject answered several questions concerning their waiting behavior and chosen waiting spot. Moreover, while watching the recorded video, subjects answered the same questions at any point where they substantially changed their walking direction. In the virtual space experiment, six experiment cases (2 cases of square × 3 types of density) were set, comprising three types of density for each of the two squares. A total of 90 samples were obtained.

12.5 Analyses of the Results of the Gaming Experiments

12.5.1 Waiting Behavior Characteristics Observed in the Virtual Space Experiment

Figure 12.9 shows the waiting spots and body directions obtained in this experiment. Looking across the entire space while sitting on a bench at the center of the square and waiting standing near a wall or tree while looking at the station were observed in this experiment; these are both real-life behaviors found in the aforementioned real space experiment, and these results support the validity of the virtual space experiment.

12.5.2 Waiting Spot Exploration Behavior

Based on the behavior traces and interviews, four types of exploration behavior up to the moment of determining the waiting spot can be identified, according to the differences in the process (Fig. 12.10).

First, (a) "Behavior approaching the target spot" and (b) "Behavior approaching the target area" are both exploration behaviors in which the subject had looked out across the square and immediately selected a waiting spot or area as their destination before starting to walk. The key difference between these two types is that in behavior (a), the subject saw a specific target spot as the destination and walked directly to it, whereas in behavior (b), the subject saw a general area as the

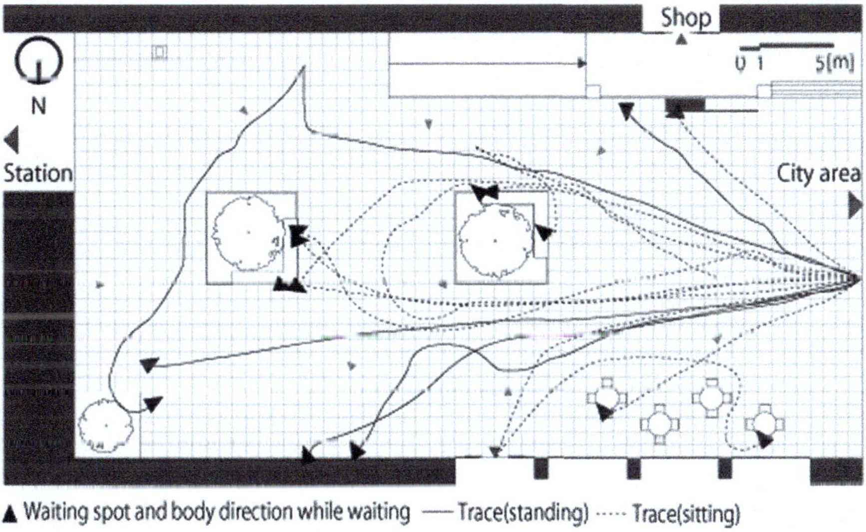

Fig. 12.9 Waiting spots and body directions while waiting (Square 1: Low density, $N = 15$)

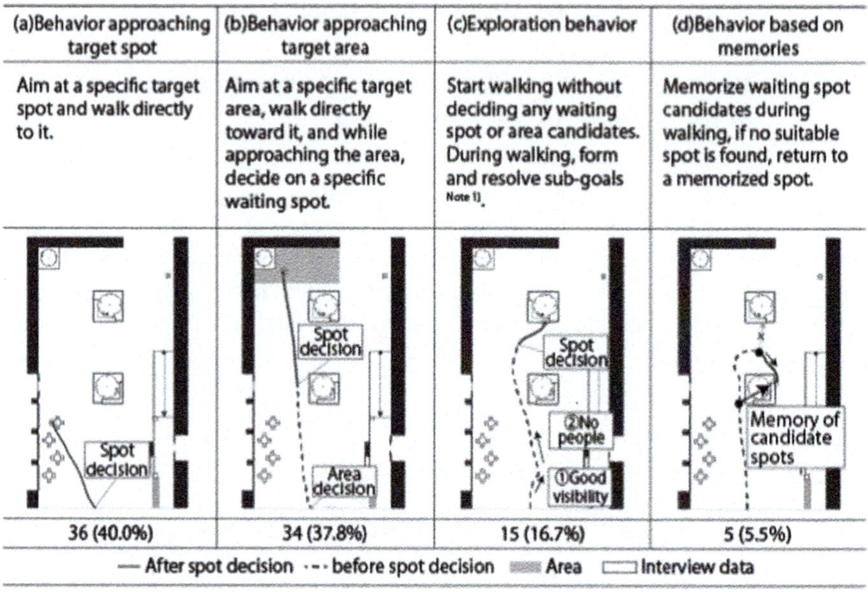

Fig. 12.10 Typical types of exploration behavior at a waiting spot ($N = 90$)

destination and walked toward it and chose a specific spot as their final destination while approaching the area. With (c) "Exploration behavior," the subject did not determine their destination until just before they decided on the waiting spot; while walking, they formed and cancelled sub-goals[1] to explore the spot. Finally, with (d) "Behavior based on memories," the subject engaged in searching and formed memories of possible spots but, upon finding no suitable spot, revisited the memorized spots and chose a final waiting spot from among them. A high ratio of subjects engaged in behaviors (a) and (b), which suggests a tendency to determine some waiting spots before starting to walk.

12.5.3 *Visual Recognition by Others: Conditions to Determine the Waiting Spot Selection*

As constraints to determine waiting spot selection, three conditions relating to visual recognition by others were used. These conditions refer to the necessary personal space to avoid visual recognition by other people. The distances were set as 3.7 m, 5.0 m, and 25.0 m, as shown in Fig. 12.11. According to Hall (1966), within 3.7 m, a person is strongly aware of the presence of other people and will generally want to

[1] An easily attainable midway destination was formed to achieve "discovery of a waiting spot" and "arrival at a waiting spot."

conditionA	Pd	Keep outside the 3.7 m visible region of other people
conditionB	Ed	Keep outside the 5.0 m visible region of other people
conditionC	Id	Keep outside the 25 m visible fovea region of other people

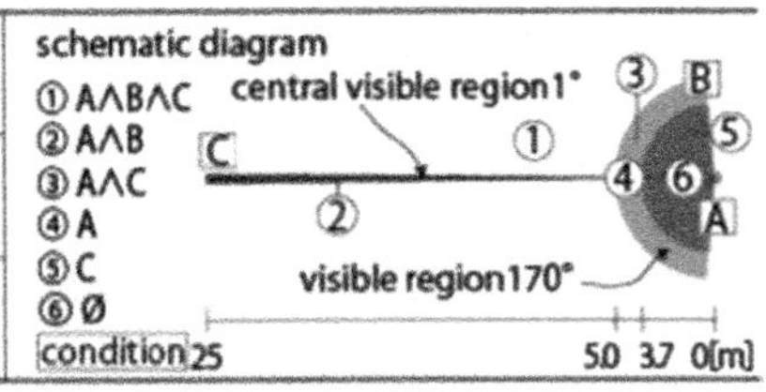

Pd: Presence distance Ed: Eye-contact distance Id: identification distance

Fig. 12.11 Conditions relating to visual recognition by others according to distances to others

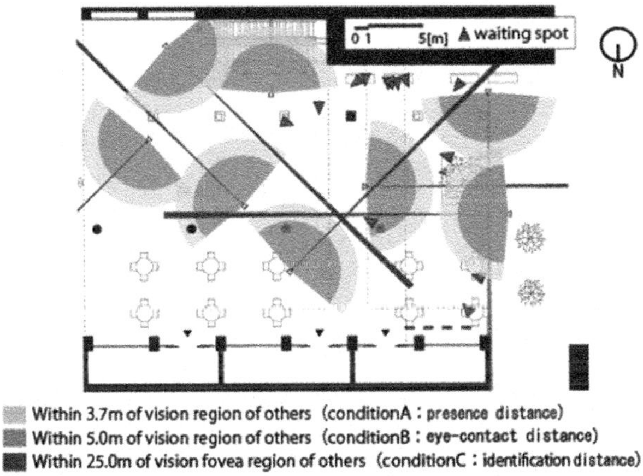

Fig. 12.12 Conditions relating to visual recognition by others and distribution of waiting spots (Square 2, low density, $N = 15$)

avoid interactions with them; in our paper, this distance was referred to as the "presence distance." Regarding visual field angles, the visible region was set as 170° and the visible fovea region[2] was set as 1°. The selection conditions of a waiting spot are illustrated on a floor plan (Fig. 12.12).

Table 12.3 shows the ratio of subjects who satisfied the conditions for each experiment case.

When these three conditions were not satisfied, a waiting spot was determined by the subject relaxing the conditions, and such broadening formed an order lattice. From the experiment sample distribution on the lattice, it is possible to infer a hierarchy of the relaxation of conditions (Fig. 12.13). In other words, when considering the waiting spot selection conditions in Square 1, Condition A had priority at low density, but Condition C had priority at high density, showing a tendency to determine a waiting spot by avoiding the gaze of other people. However, this pattern

[2]A part of the eye, located in the center of the macular lutea of the retina. The fovea is responsible for sharp central vision with an eye ground central arc of 1°.

Table 12.3 Ratio of subjects who satisfied the conditions for each experiment case (N = 90)

Experiment case	Density	Low		Medium		High		
	Square	1	2	1	2	1	2	Total
Sample		15	15	15	15	15	15	90
Number of subjects who satisfied the conditions	① A ∧ B ∧ C	9	14	7	3	2	2	37
	② A ∧ B	2	0	1	0	1	2	6
	③ A ∧ C	3	0	0	3	1	0	7
	④ A	1	0	0	1	0	1	3
	⑤ C	0	1	6	5	8	8	28
	⑥ Ø	0	0	1	3	3	2	9

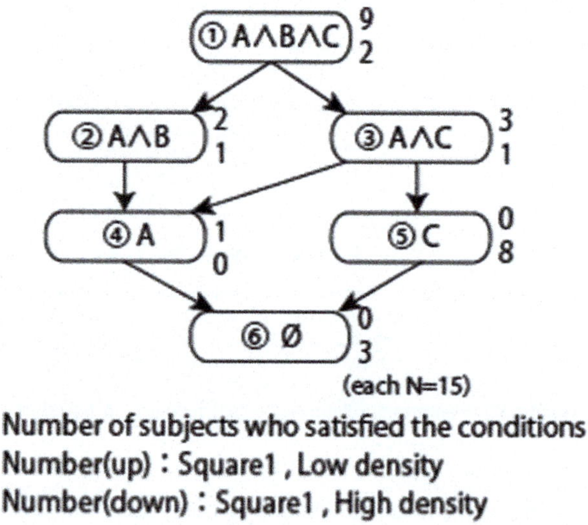

Fig. 12.13 Relaxation of determinants

is not noticeable in Square 2; a possible reason for this is that Square 2 has many chairs and other furniture items, which affected the perceived sense of distance with other people.

12.5.4 *Body Direction at the Waiting Spot*

To find key factors in determining the body direction at a waiting spot, subjects were asked questions; subsequently, factors were classified into five types (Table 12.4), which are detailed below.

Table 12.5 shows the scores of responses to decisive factors in relation to body direction for standing/sitting waiting in each experiment case.

The results clearly show higher ratios for standing at a high density and sitting at a low density. For both standing and sitting subjects, responses from high to low were ranked in the following order: discovery, vision, attraction, rejection, and front.

Table 12.4 Typical types of reasons for determining the body direction at a waiting spot

(a) Vision	To maintain good visibility	• That's because perspective and the view are good
(b) Discovery	Anticipation of the entrance from which the meeting partner comes	• In this direction I can see commercial facilities
	It is easy to find the other	• I can see a shop window display
(c) Attraction	To see specific objects and places other than entrances to the square	• I wanted to see the station
		• I wanted to turn to city area
(d) Rejection	To avoid seeing specific objects, places, and other people	• I stopped behind a café
	To avoid the gaze of other people	• I moved my eyes to avoid eye contact with the person in front of me
(e) Front	Because the body direction to some degree is naturally decided by furniture, etc	• I sat on a bench and turned to the front
Other	I really don't mind about the direction	–

(a) Vision: To maintain a clear all-round view, particularly of the front of a subject. (b) Discovery: Predicted direction from which the meeting partner will come. (c) Attraction: To look at specific objects but not seek to find the meeting partner. (d) Rejection: To avoid eye contact or specific objects. (e) Front: Any natural positioning of the body direction in response to furniture or the like

Table 12.5 Scores of responses of determining the body direction while waiting for each experiment case (multiple choices allowed $N = 90$)

(Standing)								
	Density	Low		Medium		High		
Experiment case	Square	1	2	1	2	1	2	Total
Sample		5	2	7	4	14	8	40
(a) Vision		2	1	5	0	4	3	15
(b) Discovery		2	2	1	2	9	4	20
(c) Attraction		1	0	2	0	2	3	8
(d) Rejection		0	0	0	2	2	0	4
(e) Front		1	0	1	0	2	0	4
Other		0	0	0	0	1	1	2
Response		6	3	9	4	20	11	53
(Sitting)								
	Density	Low		Medium		High		
Experiment case	Square	1	2	1	2	1	2	Total
Sample		10	13	8	11	1	7	50
(a) Vision		5	5	2	4	0	3	19
(b) Discovery		3	8	5	3	0	3	22
(c) Attraction		4	4	1	1	0	0	10
(d) Rejection		1	1	0	2	1	2	7
(e) Front		0	2	1	2	0	1	6
Other		0	0	0	0	0	0	0
Response		13	20	9	12	1	9	64

12.6 Conclusion

The authors of this study have developed a virtual space gaming experiment system to simulate waiting behaviors in a public square. Based on square research references and a real space experiment, the system was given the additional feature of eyeball expression that simulates awareness of other people's gaze.

Moreover, the present research clarified the types and characteristics of waiting behaviors in the virtual space gaming experiment, as summarized below.

First, exploration behaviors to determine waiting spots were classified into the following four types according to the difference in terms of process: "Behavior approaching the target spot," "Behavior approaching the target area," "Exploration behavior," and "Behavior based on memories."

Second, the selection of a waiting spot can be characterized by three conditions relating to visual recognition by others, which are eye contact, gaze, and personal space. It was found that there is a relaxation order on these when increasing the population density. Based on such considerations, an agent model can be implemented with crowded situations.

Finally, the factors that determine the body direction at a waiting spot were classified into the following five types: vision, discovery, attraction, rejection, and front.

References

Alexander C (1977) A pattern language: towns, buildings, construction. Oxford University Press, New York

Gehl J (2013) Cities for people. Island Press, Washington City

Hall ET (1966) The hidden dimension. Doubleday, Garden City, NY

Kaneda T, Okamoto K, Suzuki T, Tamada M (2015) An analysis of the characteristics of a proactive exploration behavior model based on three-dimensional virtual space experiments. In: The 9th International Workshop on agent-based approach in economic and social complex systems (AESCS), Bali, Indonesia, pp 93–104

Kozo Keikaku Engineering Inc (2020) artisoc https://mas.kke.co.jp/en/

Seta S, Matsumoto N, Okajima T, Kohno T, Kamiya A, Yamauchi H (1997) Development of the visual simulator directly connected with perceptual behavior. AIJ J Technol Des 4:92–96

Suzuki T, Kaneda T, Okamoto K, Tamada M (2015) Measurement and modeling of proactive exploration behavior by using a virtual space experiment system. In: The Joint Conference Proceedings of the 46th International Simulation and Gaming Association (ISAGA)/Japan Association of Simulation and Gaming (JASAG), vol 30, pp 637–650

Tsuchida H, Shigeta H (2005) Analysis on linking of place for taking a rest and waiting—a study on urban public space (Part 2). J Archit Plann AIJ 596:59–66

Yasufuku K, Abe H, Yoshida K (2006) Development of an evaluation simulator using walkthrough system and analysis of evaluation behavior from inundation into underground space. J Archit Plann AIJ 609:85–90

Yoshitomi R, Nakamura Y (1996) Behavior and cognition in the meeting place. Jpn Architect Conf Acad Lect Collect:815–816

Chapter 13
Wayfinding Simulation of Sign Layout in a Public Square by a Vision-Driven Pedestrian Agent in a Virtual Space

Yoji Yokoyama, Yotaro Maekawa, and Toshiyuki Kaneda

Abstract In this study, we conducted a sign layout survey and a pedestrian trajectory survey in Toyota City station square to explore the characteristics of pedestrian trajectories, focusing on the attributes of pedestrians. In the theory of wayfinding, it is important to consider the navigation of pedestrians. The VD18s, a vision-driven agent, was applied in a simulated space made using the game engine Unity. We perform a pedestrian simulation using a model of the Toyota City station square based on two surveys. This is a site where people move between Toyota Station (of Nagoya railroad) and Shin Toyota Station (of Aichi Loop railway). We examined the behavior of actual pedestrians, not only on weekdays but also on weekends/holidays during special events.

Based on the analysis of the results, we consider the sign arrangement of the Toyota City station square. We conclude that vision-driven agent simulations can be used as a tool to support sign planning. Not only in the square but also in the other mid-scale urban spaces, this kind of agent simulations can be effective approach for both theory and practice.

13.1 Introduction

Sign planning (in terms of sign layout and design) in public area is important for the planning and management of lively (crowded) small-to-mid-scale urban spaces. As the information on signs tends to be evaluated impulsively, the planning of signage

---

Y. Yokoyama
Toyota City Office, Nisimachi, Toyota, Aichi, Japan
e-mail: yoko402y@gmail.com

Y. Maekawa
Chubu Regional Bureau, Ministry of Land, Infrastructure, Transport and Tourism, Tokyo, Japan
e-mail: maekawa-y85ab@mlit.go.jp

T. Kaneda (✉)
Graduate School of Engineering, Nagoya Institute of Technology, Gokiso, Showa, Nagoya, Japan
e-mail: kaneda@nitech.ac.jp

T. Kaneda (ed.), *Downtown Dynamics*, Agent-Based Social Systems 16,
https://doi.org/10.1007/978-4-431-54901-7_13

needs to consider several aspects: (1) the convenience of the shortest walking route; (2) the safety of the space at busy times; (3) the comfort of the walking space itself; and (4) avoiding cognitive overload and promoting cognitive consonance with the information presented. This list clarifies that sign planning is more than a simple optimization problem.

The basic question of sign planning is how to guide navigation for pedestrian users. In other words, what positioning should planners choose to aid the navigation of a wide variety of pedestrian users? Therefore, the concept of a complex system in which the intersection between the intentionality of the various users and the intentionality of the sign planner is significant. Furthermore, if we are conscious of the theoretical scope, this can be said to be a problem involving users' spatial cognition, especially on the theme of wayfinding. Wayfinding is "the process of determining and following a path or route between an origin and destination," and involves the subprocesses of search, exploration, route following, and route planning.

Wayfinding is a theme of cognitive science (Wienner et al. 2009; Farr et al. 2012) and virtual space construction using game engines has been examined in research dealing with wayfinding and exploration using an agent-based approach (Brunnhuber et al. 2012; Chen 2012; Becker-Asano et al. 2014).

Many visitors will come to a station square with various different purposes. Therefore, at the planning stage, it is important to make a plan of the sign layout that can lead each visitor to their intended destination appropriately. This is especially true of visiting tourists, who will not have a mental model of place. For this reason, the concept of user wayfinding is important. The visitor will navigate to an intended destination according to the information on the signs, following them in order.

In this study, we research the sign layout at Toyota City station square and survey pedestrian trajectories in this square space. We then analyze the effect of the sign layout and try to reproduce the pedestrian trajectory lines using a vision-driven agent simulation.

13.2 Observation Survey at the Toyota City Station Square

13.2.1 Toyota City Station Square

Toyota City station square is a "pedestrian deck" that connects Toyota City station and Shin Toyota station on the second-floor level. This pedestrian deck has an area of 403.8 m^2 (about 20.6 m length $\times$ 19.6 m width) and there is no roof over the square except for the two arcade lanes. On the north side of this square is the T-Face department store, and on the south side stands Matsuzakaya department store.

Regarding the characteristics of the square, firstly its shape is very complicated, and the overview is poor. To make the characteristic movements of the pedestrians using the square, we carried out a survey. In cases where the view is good, there will be many pedestrians who travel between destinations by the shortest route, and few will have to search for routes. This is not the case here.

Secondly, two arcades are placed on the north and south sides, which extend to the east and west around the square; it is therefore easy to examine the characteristics of pedestrians' route choices by their trajectories.

Thirdly, there are many events that are held on a daily basis, and the impact of those events on pedestrian trajectories can also be estimated. In this square, the social experiment "Joyful Toyota Project" (Asoberu Toyota Project) has been implemented since 2015 with the purpose of encouraging citizens of Toyota city. On the day of our survey, half of the pedestrian deck was being used by food and drink businesses, and a hula dance event was also being held.

13.2.2 Researching the Sign Layout on the Square

We surveyed the characteristics of the sign layout on the square. Figure 13.1a shows a plan view of the square and its entrance, and Fig. 13.2b shows the sign layout in the

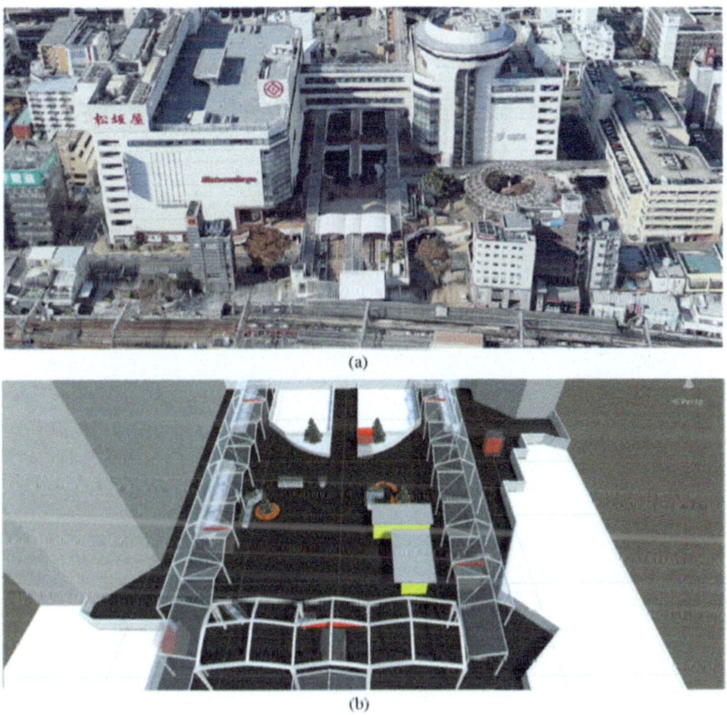

Fig. 13.1 Overhead view of the pedestrian deck in front of Toyota City station, showing (**a**) a photograph of the area and (**b**) a virtual rendition of the space

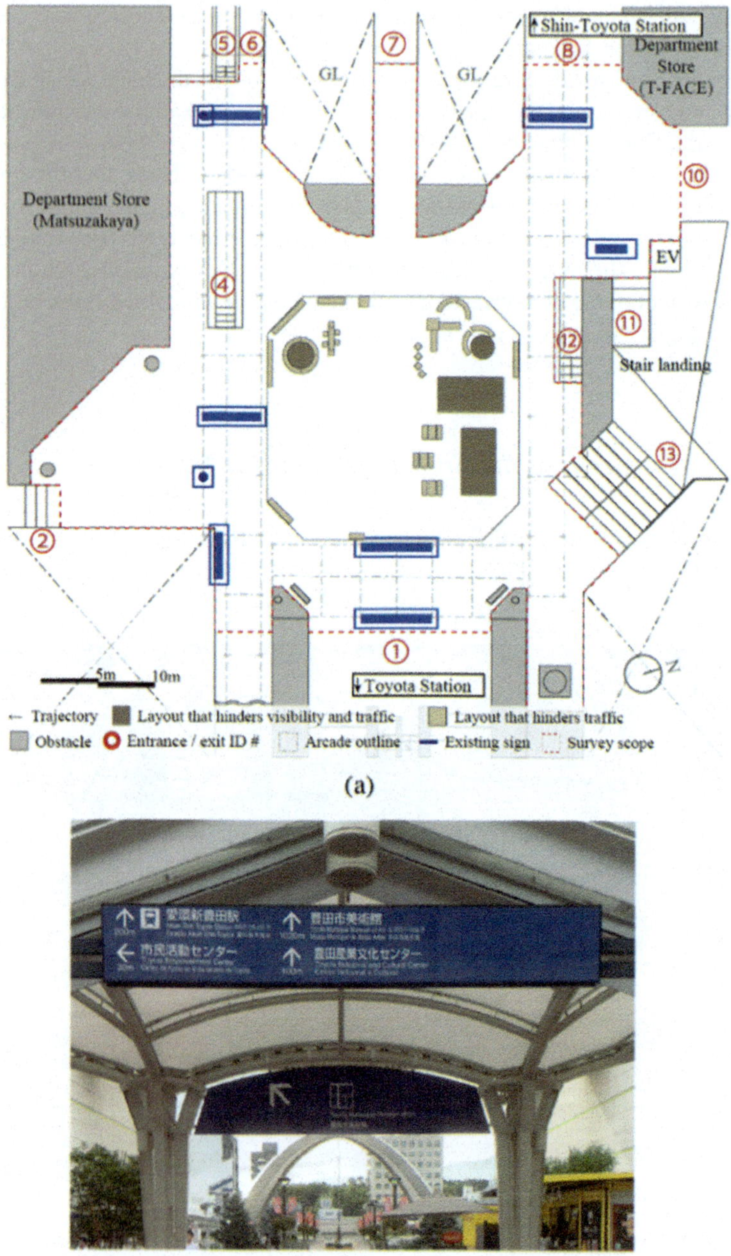

Fig. 13.2 Signs in the square, showing (**a**) a layout map and (**b**) a sample photo

square. In Fig. 13.2a, the entrances to the square are numbered, and the area inside is our survey space. In the survey of the sign layout, signs denoting place names (destinations) were our target.

Figure 13.2a shows that signs are located to indicate that pedestrians should follow the north–south path through the square. On the side lane near the Matsuzakaya department store, there are enough signs to allow smooth navigation, but on the aisle next to the T-Face department store on the north side, there are places where there are no signs along the route.

There are many signs on the south aisle to guide people to the Toyota city museum located in the southeast. Signs are generally classified into several types: transportable, installation, adhesive, hanging, and road-surface. The signs of this square are either of the installation type, on the deck ground, or the hanging type, attached to the arcade (Fig. 13.2b). Looking inside the pedestrian deck, we can recognize where signs cannot be seen because of obstacles such as plants.

13.2.3 Pedestrian Trajectory Survey

We carried out a survey to clarify the characteristics of pedestrian trajectories across Toyota City station square. An overview of this is shown in Fig. 13.3a. Between entrances 1 and 3, 1 and 9, and in the direction of the department stores, many visitors were seen. This suggests that many of the department store users came from outside the city. This is most likely because we carried out this survey on a holiday afternoon. The second largest number of visitors were seen between entrances 1 and 8 and between entrances 1 and 6. These routes are used by many pedestrians as a continuous path from Toyota City station to Shin Toyota station. On the survey day, there was an event being held, and therefore visitors tended to avoid the event space. After the event, visitors were seen in the space that had been used for the event.

13.2.4 Three Trajectory Patterns Confirmed by the Observation Survey

The results of the pedestrian trajectory survey can be thought of in correspondence with the current status of the sign layout. We can separate the transfer routes between Toyota City station and Shin Toyota station into three patterns: the left pathway, the center walkthrough, and the right pathway. These are shown in Fig. 13.4.

Although the third (right) pathway pattern is the shortest route between Toyota City station and Shin Toyota station, it should be noted that this route between entrances 1 and 8 does not have signage and can only be followed only if people are already aware of the route.

(a)

Observation Target	Pedestrians who go through the survey space	
Survey method	Observe the trajectory of the randomly selected subjects visually.	
Survey date	October 6, 2018 (Sat) 13: 00-15: 00	
Weather	Cloudy, sometimes rainy (14: 15-)	
Event	Hula Dance (13: 00-14: 20)	
Sampling Number	208 (135 at the time of the event)	

(b)

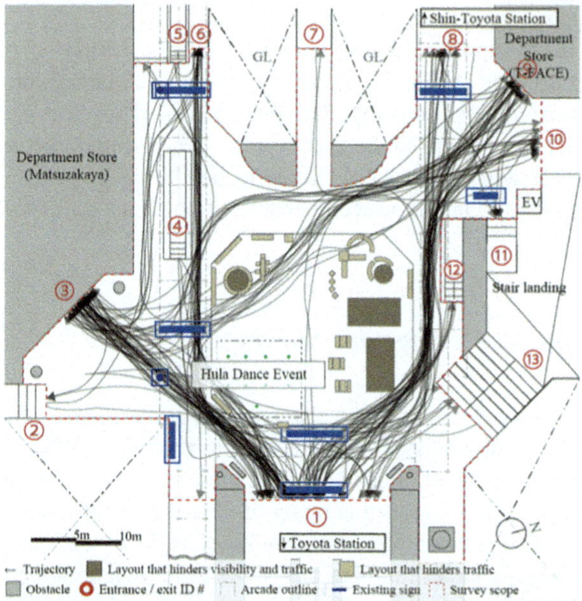

(c)

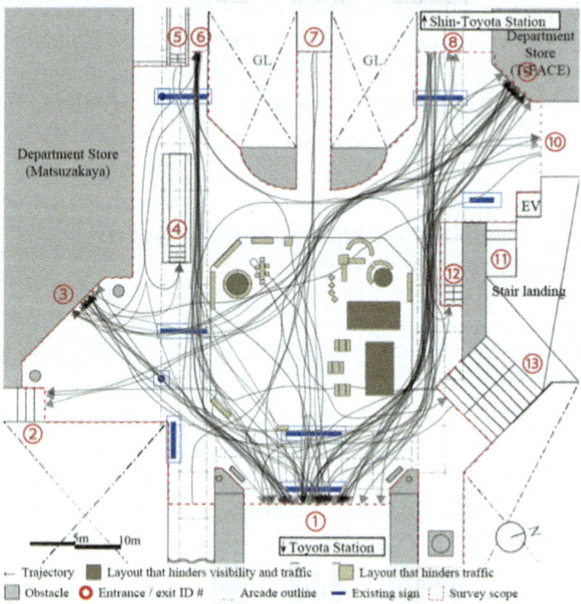

Fig. 13.3 A pedestrian trajectory survey, showing (**a**) specifications for the pedestrian observation survey, (**b**) trajectory pattern during the event, and (**c**) trajectory pattern during ordinary use

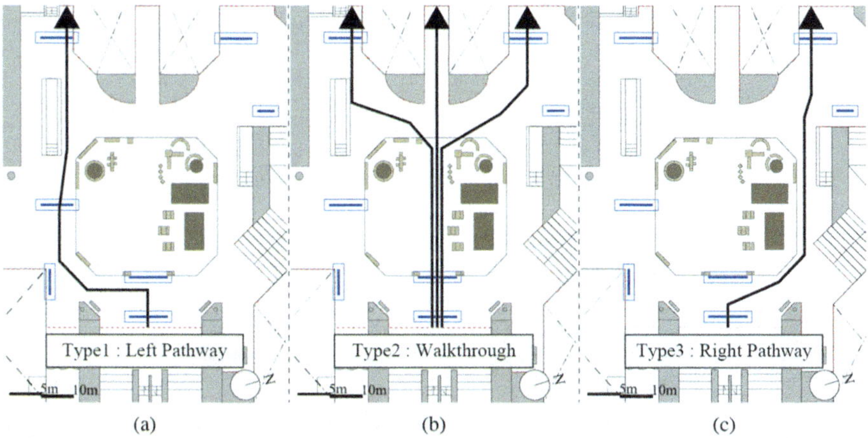

Fig. 13.4 Three pedestrian routes between Toyota City station and Shin Toyota station, showing (**a**) the left pathway, (**b**) the center walkthrough, and (**c**) the right pathway

13.3 Agent Simulation Examining the Effect of Sign Layout

13.3.1 Overview of the Vision-Driven Agent VD18s

In this study, we used the agent model VD18s. This is the first three-dimensional model from VDPA1, which was featured in Chap. 10. The expression space is three-dimensional, but the action space is two-dimensional.

The sight field of the agent model has a 170° view on the two-dimensional plane. Furthermore, there are two sight (horizontal fan-shaped) fields: one is for obstacle search from a height of 1500 mm and is used to judge the line of sight, and the other is used for obstacle search at a height of 0 mm to judge possible straight lines. Using these two sight fields means adapting a recognized avoidance rule (Moussaïd et al. 2011) in the situation "can see but cannot go straight because of an obstacle," in order to travel to the goal while avoiding obstacles. This is the largest feature of this rule. The mechanism of the VD18s model is described in detail by Kaneda et al. (2019). The walking speed was set at a constant 1.2 m/s, and decisions were made with a constant probability at intervals of 1/3 of a second.

The algorithm was modified to sequentially trace signs as intermediate destinations, and a priority level is set for each sign. When there is more than one unreached sign visible, the algorithm chooses the direction of the highest priority sign. When there are no signs visible, the algorithm chooses to move forward in the longest possible straight-line direction.

Figure 13.5 shows an overview of the decision-making process of the VD18s agent model when processing multiple signs. Figure 13.6 shows one example of a route choice in this simulation. The steps in this route are as follows.

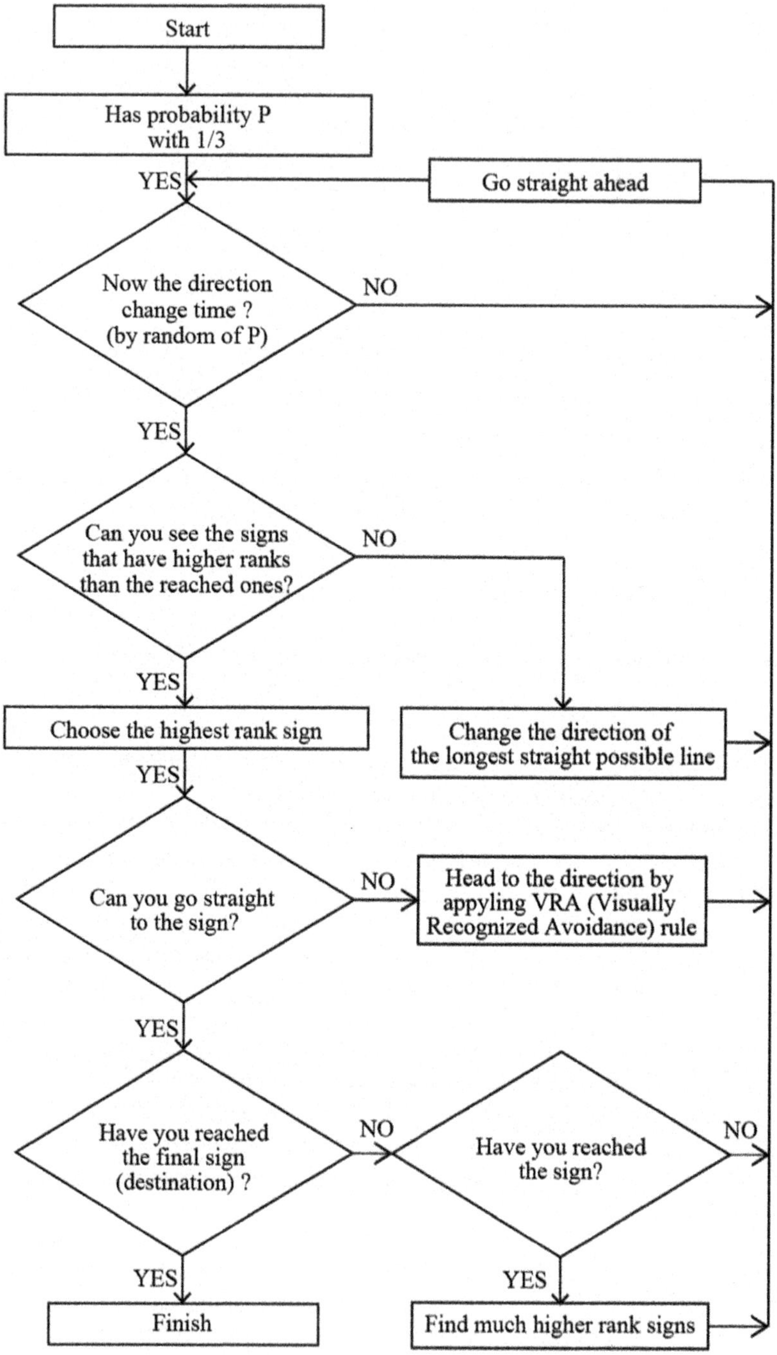

Fig. 13.5 An overview of the agent model when processing multiple signs

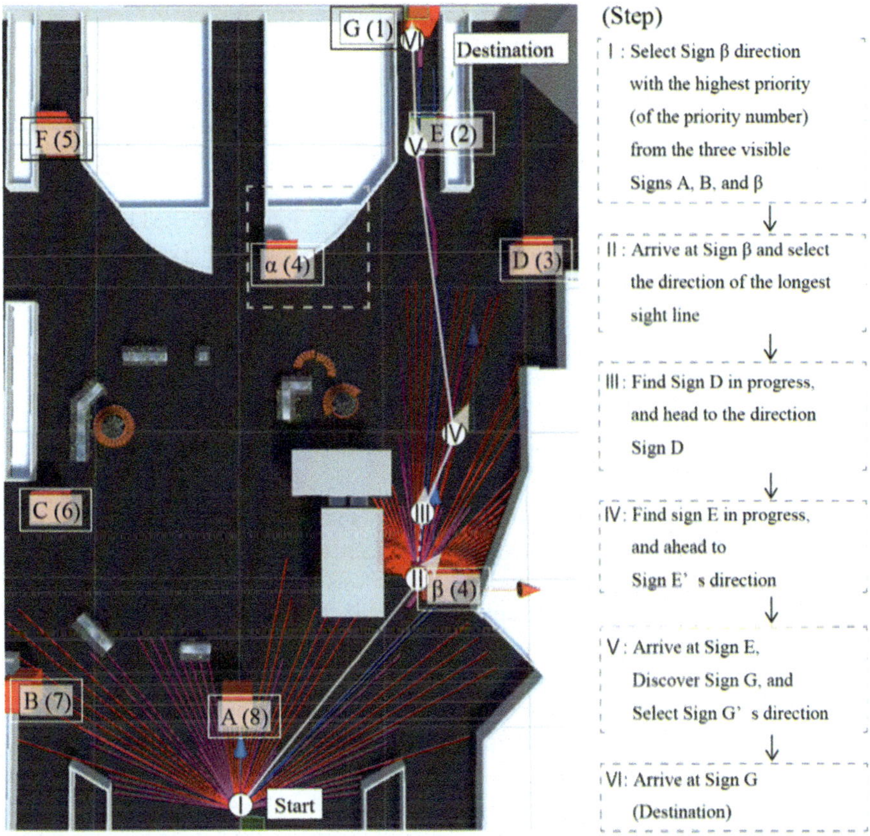

Fig. 13.6 Example of a route choice

(STEP I) The agent at the entrance of the station can see signs A (priority 8), B (priority 7), and β (priority 4), and goes in the direction of β, which is the highest priority of these three.

(STEP II) After arriving at sign β, the agent cannot see the next sign, so it goes ahead in the direction of the longest sight-segment line.

(STEP III) On the way, the agent finds sign D (priority 3), and heads towards it.

(STEP IV) On the way to sign D, sign E (priority 2) becomes visible, which has a higher priority than sign D, and the agent changes direction to go to sign E.

(STEP V) After arriving at sign E, the agent soon finds sign G (priority 1) and moves towards it.

(STEP VI) The agent arrives at sign G (destination).

13.3.2 Cases for Simulation Analysis

We created the spatial model based on actual design drawings of the square and a measurement survey. The simulation analysis aimed to consider the effect of complementary signs, and we therefore planned two complementary signs. Both signs direct people to entrance 8 and are placed so as to reduce the influence of obstacles that may obscure them. In particular, complementary sign α is located in front of the square to facilitate the trajectory inside the square. Complementary sign β is located at the north side of the square and is expected to facilitate a commuter-type trajectory (Fig. 13.7).

13.3.3 Cases 1–3: Simulation Results

We carried out simulations individually for four cases. Each was conducted 20 times, and there was no interaction between agents in this study. As initial positions for the agent, each of the four cases was placed at one of five points in front of the first exit (with a 5 m spacing between these points). Figure 13.7 shows the trajectory map that resulted from these simulations.

Figure 13.8 shows the simulation results from (a) Case 1, (b) Case 2, and (c) Case 3, and (d) shows a legend for the diagrams. In Case 1, two branches were seen; one direction (11/20) was the left pathway through sign C to sign F, and the other (9/20) was the trajectory heading towards entrance 8 (Fig. 13.8a). In the latter trajectory, on the way from sign C to sign F, sign E was found, which has a higher priority. The recognized avoidance rule was therefore activated and resulted in an S-shape turn in the direction of sign E. This trajectory is a predictable action, because plants in the square obstructed the view of sign F.

In Case 2, where complementary sign α was added, all trajectories became "center walkthrough" (Fig. 13.8b). Most agents go to sign C (only one agent goes

(a) (b)

Fig. 13.7 Plants in the square, showing (**a**) a photograph and (**b**) Unity work

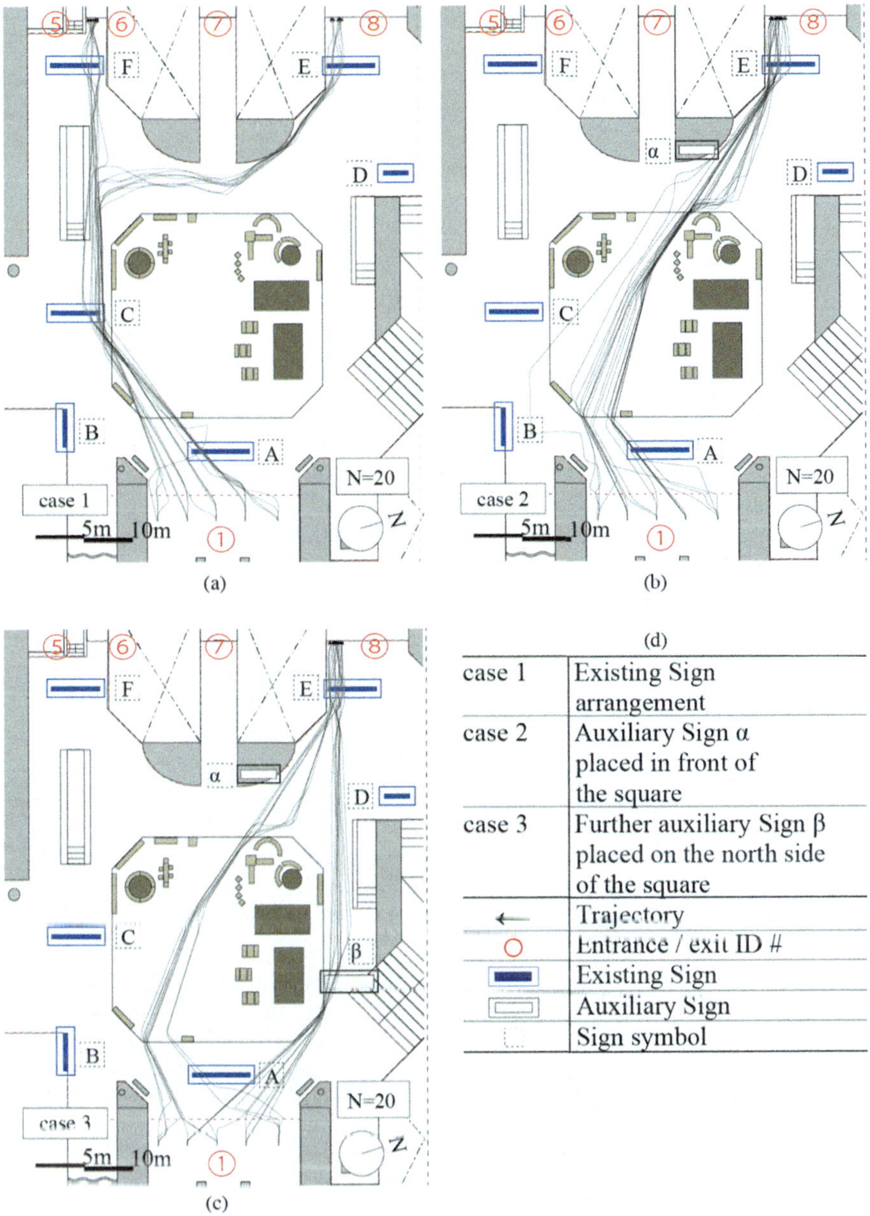

Fig. 13.8 Simulation results for cases (**a**) 1, (**b**) 2, and (**c**) 3, and (**d**) the diagram legend

to sign B) and on the way, they find complementary sign α and go through the square. They then immediately find signs D and E, and can reach sign E. Compared to Case 1, the effect of complementary sign α is clear.

In Case 3, where complementary signs α and β were included, two trajectories were seen: center walkthrough (9/20) and right pathway (11/20). The former can be explained in the same way as Case 2, and the latter goes from the start point through sign β to sign E. The ratio was almost 50/50. At the three central initial positions, both routes can be seen. Therefore, the difference can be explained as a result of the probabilistic decision-making at the choice point.

13.3.4 Simulation Result During the Event: Case 4

Case 4 reproduced the event on the date of the survey. In addition to the installation of complementary sign α as in Case 2, we arranged the event space as it was at the time of the survey (during a hula event). Unlike in Case 2, the hula event was held in the middle of the center walkthrough trajectory. Figure 13.9 shows the results from the simulation of Case 4. The obtained trajectories branched into two patterns. In one, the agent reached sign B and then found sign E. On the way to sign E, she/he also saw complementary sign α and crossed the square diagonally towards the right side. In the other pattern, the agent walked through the right side of the event space to the center of the square, depending on the principle of the longest straight possible line, and found sign F. After walking a little further, the agent then found complementary sign α and signs D and E to move towards entrance 8, exiting as in Case 2.

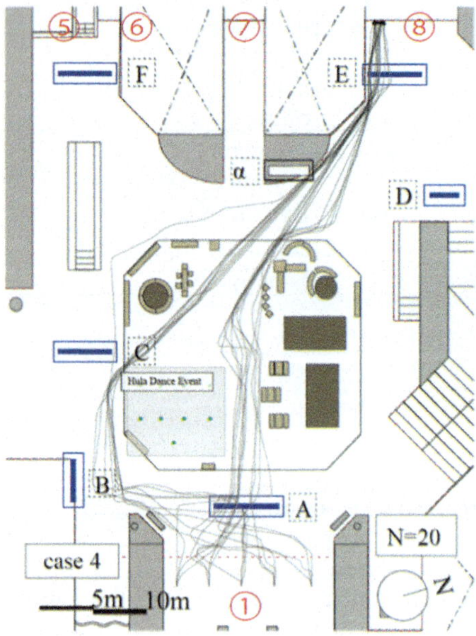

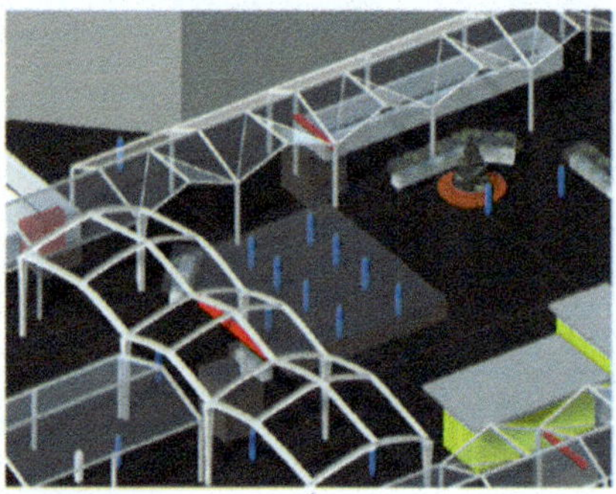

(d)

(e)

case 4	At the event (case 2)
←	Trajectory
○	Entrance / exit ID #
▬	Existing Sign
▭	Auxiliary Sign
⬚	Sign symbol

Fig. 13.9 Event case simulation (Case 4), showing (**d**) Unity work and (**e**) legend

It is notable that the approximate trajectory that pedestrians took to avoid the event space in the observation survey was reproduced by this vision-driven agent simulation. This suggests that the simulation model fits well with the observation survey data.

13.3.5 Analysis and Consideration of Simulation Results

Table 13.1 shows the aggregate results of the walking route in each case and the average walking distance. In the case of placing complementary sign α in front of the square, a visitor can be prompted to walk inside the square, and this tends to shorten the walking distance. In the case of placing complementary sign β at the north side of the square, visitors are encouraged to walk on the north side, and this enables route choices approximating the shortest distance.

13.4 Concluding Remarks

In this study, we described simulations using a model VD18s vision-driven agent based on a sign layout survey and a pedestrian trajectory survey in the Toyota City station square. In general, the VD18s model divides its route pattern according to its initial position and subtle differences in decisions regarding turning judgement.

From a series of comparative analyses, we have gained the following knowledge. The simulated event space detour pattern approximated the pattern seen in the observation survey. This is one proof of the validity of the simulation model at the Toyota City station square.

It was found that placing a complementary sign in front of the square prompts a route choice mirroring the "center walkthrough" pattern, and the sign layout on the north side of the square prompts a "right pathway" route choice. Both sign layouts are effective for shortening the walking distance of visitors transferring between Toyota City station and Shin Toyota station.

This paper discussed the status of vision-driven agent simulations for sign layout analysis in wayfinding. For greater understanding, this survey should be compared with agent model behavior through detailed analysis of observation survey results. In addition, this study indicates that agent simulations similar to this can be an effective approach for both theory and practice in mid-scale urban spaces.

In the simulations in this study, the sign layout perspective was judged by the agent from a two-dimensional field of sight at a height of 1500 mm. In the spatial model, the placement of a red rectangle expresses a sign, regardless of the actual form of the sign. In future work, the agent model will be improved by placing the sign in three dimensions so that the act of "looking up" or "looking down" can be modelled to examine the effect of different installation types and sign types, such as using adhesive signs as targets.

Table 13.1 Number of samples per trajectory group and average walking distance (m)

Case 1	Existing sign following type	$N = 11$	57.33 m	–	–	–	$1 \rightarrow 8$	$N = 9$	74.59 m
Case 2	–	–	–	Square traffic type	$N = 20$	60.94	–	–	–
Case 3	–	–	–	Square traffic type	$N = 9$	60.67 m	Commute type	$N = 11$	56.56 m
Case 4	Existing sign following type	$N = 9$	73.71 m	Square traffic type	$N = 11$	61.14 m	–	–	–

13.4.1 Visibility Is a Phenomenon "Inside" the Brain

Sign recognition models including individual recognition are "microscopic" models, which are sometimes constructed using the "inside view" based on the information processing approach of cognitive psychology from the 1970s, without consideration of the aggregate process. In examining practical problems, this tendency increases. This means that "macroscopic" modelling has not been examined because it has tried to find grounding for the aggregation process.

Visual recognition is an "inside-brain" process at the level of an individual's internal experience and is also one of the grounding methods in constructing reality. A virtual space is a mathematical model of Euclidian space, and therefore recognition in a virtual space can be one method for grounding.

13.4.2 Resource of Attention

Attention is a resource and can also be treated as a source of bounded rationality in making personal decisions. Artificial constraints for recognition, such as the probabilistic tuning conditions and the two-dimensional sight field of the agent model, should be operationally treated as parts of "attention" as a resource.

Virtual space offers to each person who has experience of mathematical model dealing with Euclidean space, reality chose to belief.

In sign-planning practice, there is careful and precise examination of visibility, but vision-driven agent simulations should be considered as a tool to reduce the workload of this process.

References

Becker-Asano C, Ruzzoli F, Hölscher C, Nebel B (2014) A multi-agent system based on Unity 4 for virtual perception and wayfinding. Transport Res Proc 2:452–455

Brunnhuber M, Schrom-Feiertag H, Luksch C, Matyus T, Hesina G (2012) Bridging the gap between visual exploration and agent based pedestrian simulation in a virtual environment. In: Proceedings of the 18th ACM symposium on virtual reality software and technology. ACM, New York, pp 9–16

Calori C, Vanden-Eynden D (2015) Signage and wayfinding design: a complete guide to creating environmental graphic design systems. Wiley, New York

Chen Q (2012) A vision driven wayfinding simulation system based on the architectural features perceived in the office environment. University of Technology, Eiindhoven

Farr AC, Kleinschmidt T, Yarlagadda P, Mengersen K (2012) Wayfinding: a simple concept, a complex process. Transplant Rev 32(6):715–743

Kaneda T, Maekawa Y, Kino T, Matsumoto Y, Susaki K (2019) Modeling and simulation of vision-driven pedestrian agent. In: Proceedings of the 12th space syntax symposium

Mehdi M, Helbing D, Theraulaz G (2011) How simple rules determine pedestrian behavior and crowd disasters. PNAS 108(17):6884–6888. https://doi.org/10.1073/pnas.1016507108. Accessed 26 Apr 2011

Moussaïd M, Helbing D, Theraulaz G (2011) How simple rules determine pedestrian behavior and crowd disasters. Proc Natl Acad Sci 108(17):6884–6888

Wienner JM, Buuchner SJ, Hoolsher C (2009) Taxonomy of human wayfinding tasks: a knowledge-based approach. Spatial Cogn Comput 9:152–165

The manufacturer's authorised representative in the EU is Springer Nature Customer Service Centre GmbH, Europaplatz 3, 69115 Heidelberg, Germany. If you have any concerns regarding our products, please contact ProductSafety@springernature.com

Printed and bound by CPI Group (UK) Ltd, Croydon, CR0 4YY
22/09/2026
02220946-0005